NEW DIRECTIONS 53

N D

New Directions in Prose and Poetry 53

Edited by J. Laughlin
with Peter Glassgold and Griselda Ohannessian

 A New Directions Book

Copyright © 1989 by New Directions Publishing Corporation
Library of Congress Catalog Card Number: 37-1751

All rights reserved. Except for brief passages quoted in a newspaper, magazine, radio, or television review, no part of this book may be reproduced in any form or by any means, electronic or mechanical, including photocopying and recording, or by any information storage and retrieval system, without permission in writing from the Publisher.

Unsolicited manuscripts must be accompanied by a self-addressed, stamped envelope and, in the case of translations, written permission from the foreign writer, publisher, or author's estate.

ACKNOWLEDGMENTS
Grateful acknowledgment is made to the editors and publishers of magazines and newspapers in which some of the material in this volume first appeared: for Laura Marello, *Quarterly West;* for Charlie Smith, *Crazyhorse* and *The Paris Review;* for Carolyn Stoloff, *Embers, Invisible City, The Malahat Review, The Nation,* and *Thunder Mountain Review.*

George Evans's "Eye Blade" was first published as a chapbook by Ric and Ann Caddel at Pig Press, Durham, England (Copyright © 1988 by George Evans) and is reprinted by permission of the author. Léon-Paul Fargue's "Three Poems," translated by Edouard Roditi from *Poésies* (© Editions Gallimard, 1963), are reprinted by permission of the publisher. "Ten Poems" by Craig Raine are reprinted by permission of the author; they are taken from his collections *The Onion, Memory, A Martian Sends a Postcard Home,* and *Rich* (Copyright © 1978, 1979, 1984 by Craig Raine), published by Faber & Faber, Ltd. The poems by "Five German Poets," translated by Rosmarie Waldrop, are published by permission of: Carl Hanser Verlag, Munich (for Heiner Bastian); Postskriptum Verlag, Hanover (for Chris Bezzels); Verlag Klaus Wagenbach, Berlin (for Elke Erb); S. Fischer Verlag, Frankfurt am Main, and the author (for Petra von Morstein); Luchterhand Literaturverlag, Darmstadt (for Friedericke Roth).

Manufactured in the United States of America
New Directions books are printed on acid-free paper.
First published clothbound (ISBN: 0-8112-1106-1) and as New Directions Paperbook 678 (ISBN 0-8112-1107-X) in 1989
Published simultaneously in Canada by Penguin Books Canada Limited

New Directions Books are published for James Laughlin
by New Directions Publishing Corporation,
80 Eighth Avenue, New York 10011

CONTENTS

Antler
    Bubble-boggled   175

George Evans
    Eye Blade   122

Léon-Paul Fargue
    Three Poems   178

Judy Gahagan
    Two Stories   9

Forrest Gander
    Two Poems   155

Paul Hoover
    The Novel   1

Nobuo Kojima
    The Black Flame   133

Laura Marello
    Catch Me Go Looking   88

Fred Muratori
    Four Poems   130

Joyce Carol Oates
    Three Stories   59

Craig Raine
    Ten Poems   65

Edouard Roditi
   Meditation on Proust's
   "Remembrance of Things Past"   6

Barry Silesky
   Six Prose-poems   171

Charlie Smith
   Three Poems   83

Carolyn Stoloff
   Six Poems   157

Rosmarie Waldrop, ed.
   Five German Poets   21

William Webb
   The Jazz Class   163

Jorge Valls
   Eight Poems   49

Notes on Contributors   183

THE NOVEL

Excerpts

PAUL HOOVER

1
Harold Ardor, perhaps Chicago's most prolific author, died Monday at Memorial Hospital. Under his own name and the pseudonym Millie Myles, Mr. Ardor wrote eight hundred paperback novels in a career that started at age fifty-two. Working in his one-room apartment in a derelict part of the city, he produced his steamy adventure and romance novels at a furious pace. One year, he wrote thirty-five, more than one every ten days. Ardor's novels weren't, however, known for literary excellence, and he never claimed to be a great writer, just a busy one. "I have maybe fifty novels that I can be proud of," he told this reporter. Despite the number of books published, admittedly for the sex market, fame and fortune escaped him. In his best year, he made fifteen thousand dollars. *Canyons of Pleasure,* for which he received two hundred fifty dollars in royalties, sold over one hundred fifty thousand copies. Mr. Ardor recited his novels into a tape recorder and scoffed at rewrites. There were no immediate survivors.

2
In the novel, where characters lingered,
running on empty was one metaphor.
Thank god for that, thought Ellen, Frank,

Johnny, Angela, Ronnie, Alex, and Kate.
Yet wisdom was never farther from truth,
and what the novelist knew
could not be held in his head.
Lightning flashed around the house,
burning shadows into the mind.
A Homer Winslow print, on a northern wall,
revealed two fishermen with nets,
faces turned from the face of the sea,
the dory frozen in space. The novelist thought,
if he ever woke up in a work of art,
he'd like to be Raymond Drizzle,
the pseudomisanthrope, bludgeoned by the butler
between dessert and brandy.

> On pg. 15, Sandy forgives him.
> 604 K used, 102 are free.
> Trotsky takes up 40 K,
> and "Flash Point" takes up 7.
> On pg. 27, the murder occurs.
> The body lies on the dining room floor
> for nearly two hundred pages.
> This proved disconcerting
> to the research department.
> We were wondering how
> you planned to resolve it,
> in prose fiction or in verse?

While passing the slaughterhouse,
the story gathered speed, and by the end
it couldn't be stopped.
Everything became it, like appetite and weather.
But excuse me, please, says the torturer's horse,
what good is all this writing,
what pleasure does it give?
For I have been one attentive
to the strictures of structure—like, look
they say, it is here,
striking the eye with a hook.

As if in a siege of language,
linear prisoners taken,
the god lay around in fragments.
This delighted the scholars of heaven.
Among acanthus and boxwood,
a boy pulls an eel from a puddle
and puts it in his mouth.
We scarcely feel we can move,
lest he discover us here,
smelling like the truth.

3
Mink patrolling their cages,
conclusive but undecided,
make perfect present tense,
as if persistence of vision,
waving a wand on light,
projected a world in space.
OK. But say that happens
and nothing leans in space to see.
No one's in the room, in mind,
or on the news. Repeatedly,
in the window, television plays
backward out of time, where
two lights seem as one
beneath the narrative tree.
In the young man's drawing
(fiction), one of his legs
disappeared in the page.
Shadows fell in every direction,
as if applied with a knife,
but they issued from our eyes
and weren't the world he'd made.
Thinking winks at this, slumbers
through like a train on which
a decision is made, to be,
or not, the thought desired.
Left-handedness and the rites

of passage? The story inside that.
The stance was saying prayers.
The shape inside the chairs.
But the house sways with sorts of windows,
being what was meant. Then music comes
from a neighboring room that represents
the world: sound of drums on water,
horns in an empty store. Perhaps
it's more affecting for being
nothing much—the way in Pinter's
plays a glass of water's the source
of tension. Fists will clench
behind it, as if behind an eye.

    Do factories in heaven
have fluorescent lights like these?
So you can look through dew
fat with subject's lease?
The object boggles the mind line:
river warden, river fever, river lightspeed.
What rhymes with "ne'er dissever"?
I don't think I'll live forever.
When the family poses for death,
the camera takes a breath.
Laid out on grass, in sight of God,
and if it glitters, it won't rot.

4
The novelist's mother said, for instance,
nothing on the day in question.
He'd come down from Chicago,
and the nurse had shown him in.
Here is your mother, she said,
while his father stood behind, ethereal as a wraith.
The nurse had done her hair, put make-up on her face,
which she never wore in life.
Beaming satisfaction. Displaying grace.
On stage. I'll have to read the work
of José Phillip Farmer, Mario Vargas Llosa,

Manuel Puig, The Ronettes, Sherelles, Chandells.
If it's going to be a fuss and hold up things,
please don't go to a bother, he said to the undertaker.
The rooms perfumed in what had been a house,
flowers all around. He had not thought death,
but of course that day two crows were in the yard.
He held her hand. Death is the process of change,
he thought, and life is radio static.
Elevator music was on his mind just then—
turgid, thin, light, and heavy.
A bowl of waxen fruit on a highly polished surface.
Things lowered into water. A blur of faces.
Someone telling a joke. "When the joke
starts to smoke," he had read.
Fear of touching a thread on his sleeve,
unraveling as a theme. Obsessive repetition,
heaviness as a theme. The sky above,
the mud below. Scatter as a theme.

MEDITATION ON PROUST'S "REMEMBRANCE OF THINGS PAST"

EDOUARD RODITI

Conceived in August in Cabourg in my paternal
Grandmother's Villa Les Violettes within walking
Distance from the Grand Hotel where Marcel, that Summer,
First began to test the magic of memory which over
The next thirteen years would slowly salvage his lost
Or wasted years, transmuting Cabourg into Balbec and Illiers
Into Combray, the following winter I was still more fish
Or newt than truly human, intent on mimicking
In my own being evolution's whole long history
While drawing my strength from my mother's blood and lungs.
Doctor Landowski, who had already attended
Marcel's mother on her death-bed now watched over mine
Through the months of her pregnancy, the first of four.
In January, when I was unconsciously striving
To become more mammal in my own semi-liquid
Small world, the river Seine rose to unremembered
Flood levels, the worst since 1746, and Marcel
Could see from his windows on the Boulevard Haussmann
A malodorous lake, while the city's whole center
Appeared quite Venetian, with its streets like canals.
Safe on the Right-Bank heights of Chaillot, my parents
Viewed this disaster from a distance, but could see
The Eiffel Tower, far on the Seine's Left Bank,

Wading with its four legs in water like the poor
Giraffe that soon died in the flooded Paris Zoo.
Fifty thousand people were left homeless while hundreds
Were drowned. My mother with other hastily recruited
Volunteers worked in a soup-kitchen. President
Fallières came and tasted their soup and praised it.
Slowly the floods subsided. I was born in June.

Of my first three years I remember mainly
The shock of sibling jealousy when a brother
Whom I was fated to survive by more than fifteen
Years was born. Chaos and confusion entered
My life a year later with the first of two
World Wars that would destroy Marcel's world and mine.
Should Marcel or I have long foreseen that friends
He or I had known would die in gas-chambers,
Or that René Gimpel, one of the three
Real-life models for Swann, would perish in a Nazi
Concentration-camp before completing his study,
So long delayed, on Vermeer, which Marcel borrowed
From their conversations in Cabourg for his portrait of Swann?
Were I now to strive to follow Marcel's lead,
My framework of years and places would prove
Too vast to be contained within fiction's frontiers
And in the end my narrator might well find himself
Attending a Dance of Death in Auschwitz, as puzzled
By the effects of privations on its guests as by those
Of age on the guests of the Guermantes party.
I've already lived half as many more years
Than Marcel lived, wandered over four continents
And met more people in far more varied
Contexts. Even the few whom Marcel had known
And I met later in Paris, Elizabeth and Antoine
Bibesco or Albert, who become Jupien
In fiction, I can belatedly call to mind
Only as I saw them, not as Marcel had known them
So much earlier.
                    Alone Doctor Landowski
Survives in my childhood memories much as Marcel

Surely knew him, with his face and hands,
A society doctor's, so heavily scented
With Houbigant soaps and lotions and his breath
Redolent of Eau de Bottot whenever
He was summoned to examine me for my asthma
Or else for the mysterious losses of consciousness
That proved forty years later, after the recent
Invention of the EEG, to have always been
Mildly epileptic.
                Like Marcel I was indeed
A sickly child, and I'm often surprised
At having survived so many much healthier friends
Who still haunt me in the chaotic world of dreams
And unconscious memories that consistently confuse
Reality with wishful thinking in time and place.
Perhaps I should now record only my dreams
Rather than strive to unravel the skein
Of memories that appear so much less meaningful.

TWO STORIES

JUDY GAHAGAN

GABRIELE D'ANNUNZIO'S ASTONISHING POWER
OVER WOMEN

It is said that women were so mesmerized by Gabriele D'Annunzio, the famous writer, that they would abandon husband, family, children, home, possessions, friends, social position, interests, careers, and fame, solely for one night with him. . . .

In this way it is thought that he slept with around a quarter of a million women.

He was, apparently, small, ugly, and prematurely bald.

"Oh, youth, alas, the crown that withers on my empty brow," he writes in a poem, obviously inspired by his incipient baldness.

Gabriele D'Annunzio preferred a particular type of women. He liked best women who were young, astonishingly beautiful, of a distinctive and refined elegance and impeccable lineage, who would share his thirst for erotic experiences of all kinds, and who would dedicate themselves to his career while having remarkable talents of their own, which they would abandon without demur to follow him wherever his frequently changing whims might take him. For example, the brilliant young opera singer Contessa Marialuzza Teresina Mandrin, a beauty, with lustrous black curls, and huge blue eyes, and skin like pearls, dressed always in black moiré or lace, gave up singing to spend months copying out, in longhand, a new work by Gabriele D'Annunzio, a drama on a classical theme, lasting six hours and using 49,780 different words. While waiting for her to finish this task in his lodgings in Rome, he sends a telegram from a friend's yacht somewhere in the Aegean: "Now I can think of nothing but sailing. I need action, adventure, danger, discovery. The sea, the sea. Your beloved Gabriele."

All the women who had what we would call today "a relationship" with Gabriele D'Annunzio came to very unhappy ends. They ended up abandoned and alone, penniless, sick or mad, careers truncated and spoiled, addicted to drugs or drink, their faces and figures ravaged by grief, despised by all, enduring the long agony of hearing about his latest conquest in the glittering salons of Europe, for he was not one to keep his affairs quiet.

So we must ask ourselves the question, "What was the secret of Gabriele D'Annunzio's remarkable power over women?" In order to address this question we have to consider the evidence that has been available to scholars from an examination of letters and diaries kept by those who knew him. Many scholars have emerged from their study of D'Annunzio with very similar conclusions. It seems more or less certain that the secret lay in D'Annunzio's *gift for words*. His eloquence spellbound every woman who came into his presence. As I have already remarked, in one work alone he used 49,780 different words, which gives you some idea of his vocabulary. And that was just Italian. He spoke French as well.

"He certainly talked a lot," said the exquisite Marchesa della Margarita.

"You could hardly get a word in edgeways," confessed a young parlormaid pressed into his service during his declining years.

Nevertheless, occasionally his courtship of a woman ran into difficulties. These were often caused by practical circumstances: maybe the irritating presence of a husband, or one of his own previous mistresses, or more frequently, the complete lack of any accommodation, having been evicted from his previous lodgings and all his possessions auctioned off. In such cases the woman might not succumb and agree immediately to accompany him to the dusty torn couch of some penniless fellow artist. On encountering difficulties of this sort, he would literally deluge the woman with telegrams and letters extolling her beauty and the urgency of his desire for her. If he could borrow money from a current companion, he would follow this up with great bouquets of exotic flowers and baskets of fruits and chocolates. Then she would succumb like all the others.

All his life D'Annunzio was haunted by money problems. At one stage he was supposed to be supporting: his wife, with whom he'd eloped aged twenty, and whom he'd left, three children and three years later; his next mistress, by whom he'd had another two children; his profligate father and dying mother. These responsibilities placed a severe strain on his income, which in itself was rather unstable. The problem was compounded by his expensive tastes. Aside from the generosity of his courtship habits, he bought a lot of clothes. At one stage he had 252 shirts and several antique hats, one alleged to have belonged to Napoleon, whom it is believed he greatly admired.

Eyewitnesses of the period tell us that he had exquisitely refined tastes. He always had a plentiful supply of Russian cigarettes, the finest wines, the most exclusive French perfumes, and "biscotti peekfreen." . . .

He found it difficult to save or plan ahead.

Once, with the generous advance from his publisher, which might have alleviated the sufferings of at least one or two of his dependents, he immediately bought three race horses and a splendid Newfoundland dog.

"Obviously he had to keep up appearances," said the Baroness Haghorn Vlaeminck, an auburn-haired hunting companion. And she was right. He was often invited to address audiences consisting of: heads of state; the highest ranks of the armed forces, the judiciary and the church; renowned men of letters from the capitals of Europe; and the most celebrated personalities from the world of the arts. When he finished speaking (according to his own account in an autobiography thinly disguised as a novel) they would be silent . . . stunned by the power of his poetic vision. Then they would break into tumultuous applause. The next day all the newspapers across Europe would fill their front pages with his speech.

Some people objected to D'Annunzio's love affairs. They said that he was a danger to women. Others argued that he intended no harm:
"I am intoxicated by the eternal feminine" (*la femina eterna*), he writes to a friend.

It sometimes happened that, tormented by jealousy of some new beauty, a mistress would write to him:
"Don't bother coming back."

Then he would be lacerated with grief. Immediately he would deluge the woman with telegrams and letters, extolling her beauty and the urgency of his desire for her. Maybe flowers too, if he could afford them. Sometimes he would travel the length of Italy to, as he put it, "renew his love." He made a particular effort if there was some hint of his mistress exciting the interest of another man, or of the possibility of being taken back and forgiven by husband and family. In every case the mistress would relent, recalling the early days of their passion, swearing her eternal love, and asking only that D'Annunzio should renounce all other women and remain faithful from then on. Shortly after this, he would terminate the relationship completely.

Some have maintained that D'Annunzio's power over women continued into old age. Others, however, have reported that towards the end, his courtship ran into difficulties and his style deteriorated. Indeed it has been alleged by critics that he would jump on any famous beauty, left alone with him, and struggle without ceremony to remove her clothes. Despite such reports it is definitely true that in his mid-seventies, toothless and dribbling, his skull covered in eczema, yellow from his addiction to cocaine, one-eyed (he'd lost the other while dropping patriotic poems on the Austrians from an aeroplane during the war), he sustained a passionate affair with a nineteen-year-old blonde German waitress.

"Together we have reached the wildest intoxications of love," he wrote to her. It is not known how she replied.

Shortly before his death she went to Berlin to live with Ribbentrop.

Mussolini gave him a warship, which he kept moored at the bottom of his garden on Lake Garda.

When he died Mussolini ordered a twenty-one-gun salute from the warship.

LANDSCAPE WITH YELLOW HOUSE

I saw the house, or one like it, for the first time over thirty years ago. I was ten years old then. It was in a picture my father had painted for a new friend of his. In the center of the picture there were people on a terrace all looking in different directions, surrounded by stone vases filled with oleanders, flowering trees, and beyond, the cypresses and mountains. From one side of the terrace a stone staircase waited in the sun for someone to go down it. And on the hillside at the extreme right-hand edge of the picture there was the yellow house with green shutters—all closed.

I haven't told him yet. I'll have to soon, for one thing they'll put up a SOLD notice, and he'll see it when he goes out to buy his paper. He's got used, these last years, to my calling in every day. We spend a couple of hours drinking tea and talking about his pictures. He still paints, but only very modest things these days. His wife encourages him, for it keeps him active and reasonably cheerful and undemanding. Not that she is there much. She has her own life and family. She looks after him in her fashion—but she is so much younger than he is, and she's got bored with him. Especially with his melancholy. He is like the weather in the mountains, when the clouds come down for days on end. There are days when he broods, slipping down into some dark region of his own, where he remembers his painting, his brief fame, then the years of invisibility and her growing indifference—and probably his other, his real family too. Those days of brooding are very long. When you are old all days are very long. He slips down into the dark, and if you come upon him suddenly he looks flustered and angry, like a cat with its ears on one side and tail twitching. And down in that underground territory he suffers. I am the only person who can winch him up to daylight again, and make him laugh; refurbish his sense of pride, for I am an artist too, and I can discuss his work with him as if it still mattered, as if we were part of some artistic circle, as if he were still important. As if I had forgiven him.

I was told to mind them, give them their dinner, and make sure Lala didn't put sand in her mouth. They set us down behind a small grubby sand dune, held together with coarse grass, to keep us out of the spiteful wind. Lala and Colin dug and scrabbled peaceably about in the sand. Mannie's nose was running, and she waited red-fingered for someone to tell her how to play. I watched my parents. They were walking beyond the breakwater at that moment, unaccountably arm in arm, as they probably had done before I'd been born. They were walking along like real couples do on promenades—like couples all over the world, in the choral dance of couples. They looked as if they had been gift-wrapped—she fair, he dark. Just for a moment I thought something impossible had happened and they were going to make it up, and he was going to get a regular job, and we were all going to go on living together, and *she* would disappear. No, I didn't think at all. My throat was filled with impossible insatiable longing. I knew that they were going to part. Mannie, red-handed and runny-nosed, would be sundered from Colin. Colin was to go with them. He was too much of a handful. He didn't know this. I did. Being older than the others, I was something between a servant and a relative, so I heard things. But I never said anything, or anything that anyone ever seemed to hear, for I was quiet and invisible.

They were still walking together. My father was waving his hands about. And definitely telling lies. I can't imagine what my mother would have been saying to him. He would definitely have been telling lies at that moment, for he had another woman. She was very well off, well educated, with thin, cruel-looking ankles and a posh voice. She did not live on a council estate. She lived in an old cottage with beams and chintz armchairs and owned an art gallery and an antique shop. She had taken an interest in my father's painting, and in my father. She had matching, unchipped cups and proper mealtimes. She was very young and clever, though not pretty, like my mother.

They'd turned and were walking back now like a jury with its verdict. I wondered why they'd brought us to this place as if for a treat, when it was for the planning of an amputation. Not that I liked the place. It was raw, untreated seaside, with unfinished-looking bungalows and a rusty funfair. But it had a flat sandy beach for the little ones. I would have chosen cliffs and a castle and fishing boats and caves. But we always came here. And I knew this outing today was something to do with "talking things over." As they came nearer I could see the look of petulant revenge my father was arranging on his face. The others were too young to notice. He was preparing it for me, where it would have the most impact. Like a child, he wanted someone else to suffer in his guilty dilemma.

The first thing I ever noticed here was the shutters. How they were closed, and the inner life of the house a proper secret. But how in the mornings girls with shining hair and solid-looking women would fling them open with a crash and call out to people in the street. My yellow house with green closed shutters is on the end of the village square. At the other end of the street are the mountains. I look forward to opening the shutters in the morning and the sun streaming into my room like a marching band, and the scent of figs and oregano. I suppose once I know people here, I too will open the shutters with a crash, and call out to passers-by. Maybe living here my voice will get louder.

After he left I was always ashamed at school and spoke very quietly, and the teachers were always saying "Speak up, speak up." Then

Miss Cottle devised a very terrible singing lesson. Miss Cottle would sing, "Who has the penny?" and point unmistakably at one child. Fifty-two eyes would aim like guns, and you had to sing, "I have the penny." My voice was small as a threadworm. "Louder," sang Miss Cottle. I sang again in my threadworm voice. The children all laughed. "Louder," sang Miss Cottle. And I blushed as if my handsome lying father, his cruel-ankled patron, and my mother lying on the torn sofa were passing through the classroom on a float.

Though I love music I never sing. Well, not like the people around here in this village sing. They sing so loudly, as if they had never been ashamed of anything in their lives.

I'd forgotten to give the children their lunch—a bag of crumbling, soggy sandwiches, processed cheese, and some scraps of cold meat, prepared by my mother. There were one and one-half sandwiches each. My father might buy us ice cream later, but he would only buy three. I hoped Mannie would leave her sandwich, she usually did, and then I would fall on it. I was always ravenous. Allowed out in the lane at home, with Lala and Mannie in the pram, I would stuff rosehips, and clover, and hawthorn berries, which we called bread and cheese and blackberries, even though, unwashed, they had maggots in them. I would stuff all this into my mouth shamelessly. And at home I stole food, I stole sugar lumps, dried prunes, the corners of new loaves, pieces of cheese. I would find my book and read it greedily and stuff the food without even tasting it.

He sat down awkwardly, glaring at me from under his long lashes. My mother's face looked as if it had been pumped up, her eyes were glassy and dotty. Tonight when the little ones were in bed the fighting would start and there would be such terrors. . . . I and the neighbors would hear it all. The day of my father's departure was approaching like the day of an operation without anaesthetic. He'd take off with his paints and pictures and brushes and oil—and the picture with the yellow house.

Never again would I see that magic hatching of a picture from the blobs and strokes and smudges, as I had watched him, hour after hour, in the back bedroom. For even when I was as small as Lala I could spend hours watching a picture becoming. It always happened so suddenly—the change from chaos to a picture—

and it seemed to me that he was like a genie and I was his special assistant, well, he always made out I was. Sometimes I would join in . . . "Make that a horse . . . make that a bush . . . make that a cloud." Sometimes he would give me a palette and piece of canvas to make my own picture, on the bench next to him. But my pictures never impressed him. He never bothered as other adults did to say, "Well, now, look at that, isn't that splendid." I had the feeling I could never please him. If the picture was poor it was unworthy of his attention, and if it should ever be good he would feel himself rivaled in some way. The yellow house had been in the picture right from the beginning, before I'd even seen it—it had been waiting for the picture to happen around it . . . just as it has been waiting for me all this time.

Mannie ate her sandwich and asked for more. I looked at my father, and he got up and came back with three ice creams, and Mannie and I shared. I golloped my half like a mongrel dog.

The yellow house has a large kitchen with an enormous open fireplace. The floor is made of bricks set in herringbone pattern. There is a wide, shallow sink under the deep-set window looking across the vineyard, and a huge wooden table. In this kitchen I shall eat as the people here eat. All food here is cooked, and no one takes as much as a glass of water between meals. The bread is wrapped in a damp cloth and put in a basket, and everything else edible is put away in tins and dark cupboards in rooms darkened by closed shutters. Food is only found in the kitchen. No chocolate or biscuits in bed. No chips on bleak windy corners: no crumbs in chairs; no stained mats; no cups half-filled with dead cold tea; no banana skins. When I first came here and stayed with my friends I remember the hunger of the long hours between lunch at midday and dinner at 8.30. And it made me remember the hunger in the lane and picking berries, and the hunger in the cold dirty house and stealing lumps of fresh bread. But gradually I came, as my friends here do, to fill the long afternoons with mending, writing letters, reading, and for me, painting. And I began to forget about hunger. Then would come the measured ceremonious dinner of simple food, and talking about the cooking of the food. That is how it will be in my house. Eating will become as a fine harvest, as a tilled field, as a well-run house.

At last I am moving to my yellow house with shutters, and I won't be back in this country for months at a time. The move has been in the air for so long. Having been quite successful, I can get enough commissions to live and work wherever I like. I have had much more luck than he had. Even though I never found a patron to support me as he did. It's not so far away, but I won't be on hand for our long talks about painting, to look at the bits of work he still does and encourage it, to recall the illusory past of his talent. Of course when I am gone, gone will be his last natural link with his family. His real one. For Colin left at sixteen never to return or even write a letter. He never saw the little ones again. They grew up and married happily years ago and forgot him. She lost interest in his painting, all painting years ago, and moved onto other projects. Her future lies elsewhere. She is so much younger than he is, you see.

My mother used to spy on them. She would wait until dark and creep up to the cottage window, and if the curtains were not properly drawn she would be able to watch them. Usually they were not properly drawn. The inner life of the house was not sacred, as it is when you have shutters. So she would watch them and then report back to me with mad glee. The first time on her reconnaissance she saw the picture of the terrace and the yellow house, standing against the sitting room wall in the old cottage, even though he wasn't actually there on that occasion. The picture was proof. The next time he was there and she got even more proof. She wanted me as an ally, but I didn't want to be.

When the final row came, it came at night, and as usual I heard it all, for our skinny walls had no respect for place, time, or person. I knew it all already by heart anyway. All the pogroms had happened at night. It was at night my drunken uncle, with his big boiled face and tough as a Boer, had stood over my father, as he struggled up from the floor, blood seeping out of his mouth. It was at night they would haul me from my drenching bed, with accusation and counteraccusation and slaps and threats. It was at night that my father, shouting and sobbing, and my mother, viciously and crazily calm, took their leave of one another. Then I heard hours of bumping and dragging as he took the rest of his pictures away and his small cardboard case and Colin. And I heard her car outside. And he was gone.

My mother did not get up for four days. She lay, looking gray and frightful and refusing to open her eyes or speak. Mannie cried incessantly for Colin. She was not to see Colin again for ten years. Later my mother and her friends said, "Good riddance—let her have him—she'll soon know her mistake." I knew better than to say anything. I had nothing to say. I was a dark sulky witness. And so I grew up invisible, ashamed, and hungry, until I learned how to paint.

Eventually I was invited to visit their new life together. They moved to another town. And she paraded his talent, and ran the gallery, and found sponsors and organized exhibitions, so for a while he was a full-time proper painter. And she loved it. She dressed in turbans and drapes and showed him off to her friends, for he was very handsome. And they filled their house with people you couldn't trust and had parties. And I appeared, darkly, making minimal claims, with my belongings in a paper bag, and bringing news of the mess on the council estate, and how the allowance the two of them made was being spent. I never saw much of Colin. He went out with his friends when I came. And so I ingratiated myself with them and helped them to soothe their guilt. And he gave me back the picture of the yellow house.

But this period of glory did not last long. It was a flash in the pan. The commissions dried up. The notices became critical or absent altogether. The friends, contacts, and sponsors always seemed to be elsewhere. And he was reduced to helping out with the gallery, and when that went, the shop. He stopped asking about the little ones. But he was always interested in my studies, and the art school, and when I began to work independently he was very proud, not rivalrous at all. It seemed to make up for the fading of his own inspiration. And she moved on to other things. Not that I bothered finding what these were. We had little in common. The old days, like the day at the seaside, had been forgotten.

SOLD says the board. I didn't tell them before. I let them see for themselves. By next week my furniture will be in store. By next month I will be in the house which I first saw in the picture my father painted for his new friend, yellow with green shutters, all closed. It's a pity my mother never lived to see this house. She would have loved its thick walls and clean brick floors and dark shuttered rooms and huge cupboards full of linen. For her own house, the house where she and I and Lala and Mannie had lived, had been cold and squalid.

The large room at the top has an open gallery, and there I shall work. The dining room is a huge white arched room with arched windows with inside shutters. My friends and I will have dinner together, laughing and drinking wine, and we will sit outside in the gallery and watch the stars shift to the other side of the tree, and listen to the crickets, and smoke and talk about painting. And in the morning I shall throw open the shutters and call out loudly to whoever is going past.

And I shall have a huge party outside, like a picnic, with Chinese lanterns in the olive trees, and music, and the big kitchen table will be carried outside and loaded with meat and fish and olives and fresh cheeses and pies and fruits and cakes. And to this party I shall invite Colin and Lala and Mannie and their spouses and children.

I don't think he ever believed I would do it. I had talked about it for so long. I think when I talked about the problems with the contract and the price and the delay he had had a moment, as I had had on the beach, of thinking that somehow everything could be miraculously restored, and that somehow I would stay. He might have believed in a last minute reprieve. But when he saw the board he would have known it was not to be. And now the day of my departure is approaching like the day of an operation without anaesthetic. He will be by himself, as I was, when the little ones cried. And he will slide down into his black river, and he will have to pull himself out. But I'll leave him the picture of the yellow house.

FIVE GERMAN POETS

HEINER BASTIAN · CHRIS BEZZEL · ELKE ERB ·
PETRA VON MORSTEIN · FRIEDERIKE ROTH

Selected and translated by Rosmarie Waldrop

SIX POEMS: HEINER BASTIAN

VIEWS

something in us that you look at
like an "internal quality"

something invisible stable
remote like a suspicion

there might be only one order

"one language or none"

a summer day in 1887
the *bridge in asnières*

a bridge
but you can't find the town
it's invisible as if it never existed

on the picture the bridge stands out
bright clear

a summer day . . .

you see a lady and a gentleman
walking along the river
in long black clothes
a train slowly crosses the bridge

I stood a long time in front of this picture

the strollers held their step
I watched the train
but now it seemed to stand still

the same painting
all movement mysteriously gone

now there is no sense trying
your eyes won't find it
movement is so remote it seems
the same as with the town of asnières

a year which again and again follows another

[(do you remember the conversation) . . . the *bridge in asnières*]

"one language or none"

a cityscape in winter

clanging air
rubs your eyes clear
and any day which starts to be like another
is already lost

do you remember . . . the *bridge in asnières*

how strange that you know
what I am talking about
that you understand it all

[Author's note:
fourth stanza: "one language or none" is a quote from lars gustafsson's *die maschinen,* munich, 1967, p. 36.
fifth stanza: *bridge in asnières* is the title of a painting by émile bernard, 1887. it is in the permanent collection of the museum of modern art in new york.]

PREPARATIONS IN LENINGRAD

(for lars gustafsson)

in the summer of that same year
the observations were nearly completed

a casual note
with few distinguishing marks
found on an afternoon
in leningrad
remains the only record

any other details
an unknown story
the expected arrival of a gentleman
just point more definitely
to the end of summer

you don't want anything
you could picture something
much more mysterious
than the vanishing of distant years

from here

any apparition is like
certain other apparitions

the events
lead you
up a gently rising road

it is winter
strange light
between the houses

what you're looking for
is
a quite different
much more distant story
at the end of the open road

[Author's note: in the winter of 1966/67, in a moscow museum, I saw an etching of the city of leningrad from an old collection. the print showed a system of evenly rising and falling streets which cannot have corresponded to the image of the city at any time. the picture seemed unreal. it was the occasion for this poem.]

BEFORE MARGINS

this is the climax of the story

it is noon

strange look
this town is taking on

restless movements continuous
ever new
approaching

a giant image
before it suddenly
fades from sight

all at once
you turn up here
and if you wanted to recognize yourself
in all the enterprises
then the principle
is probably lost

as if nobody had noticed
out there on the still water
a ship
announces its booming arrival

on the pier silence

given the deserted square
the observations reach
deeper and deeper
into the land

SIGHTSEEING SOME TIME BACK

but east of africa
we find the indian ocean

you stayed day after day and yet you've never
been there

similar topographies
occur near the pacific
I've seen them myself

one winter though trips had long taken me elsewhere

I lived there
always
in wrong contexts

those coast formations . . .
a chance geography

the efforts of huge masses of water
are like those impatient visits
and you're like them
there's no proof

a series of correspondences

it could be continued
in ever new stories

but if you've never been around
anywhere there
don't read on
you wouldn't understand

A STRANGE STORY OF ROMAN K

a summer evening that strikes us
throws light on the moving details
of the case z

the blue of the sky
it's so strangely close
that even the houses the distant other city
look at us like transparent architecture

in the continuing story of roman k
this is the point
when the sun reaches its zenith

we see z follow
apparently without purpose
the line of a beach
before he fades in that story
into a fainter and fainter object
suddenly invisible
gone forever

this is in june
near a mexican city

z is unknown here
so that the story
seems all made up

on a summer day

when the place described is any place
and the gentleman I refer to
everybody or nobody

language will run through your fingers
like fine sand
which you hold
for the short space of understanding

OBSERVATIONS IN THE AERIAL OCEAN

this is a road which continues
evenly endlessly

you could also say

"this road has no end"

"as far as I'm concerned it's endless"

"I won't think of an end"

it's strange
when you walk here
your ways lose their unreal quality
it's nearly
like following up certain descriptions
until you get to a distance
where the end
is forever out of sight

and there begins the realm
of suppositions

but let me tell you
there is no such road

what you see is a splendid atmospheric phenomenon
the memory of a conversation
on a cold winter day
when we walked and walked
and walked

a straight road
light falls from a great height

until it gets weaker and weaker
every year
and the words
have told you everything

it is strange

if memory belongs
with the logic of language
the secret of the conditions
is altogether unlikely

a straight road

and no matter where you start
it seems ever in vain

here where you walked just a moment ago
the words have told you everything

you're supposed to accept that
you're supposed to accept that

CHRIS BEZZEL: FIVE PROSE-POEMS

1
white-forsaken stony. the ferries run every once in a globe. onion quiche, they say. mother's a fake. in the woods, tests of courage. moon on the daffodils, rust in the with-pleasure-fills. untopped, the foaming sea. in vienna, at all streetcar stops, roman candles. mass arrests in nürnberg clarify the situation. in snow, mussels are mussels in snow. white-forsaken stony i pass. the pus paid up. the ladder rusted. in my flour bin i'm king. today is january.

2
this direct way of writing gives you directions for bluish relationships in bomb shelters. kitzingen am main is a world. the world is everything that is the case. let's case susanna. on february 23, 1945, umbilical cord cut. in the brick factory, the economic atomic miracle goes to rot. healthy sentences knocked bloody, and susanna, the rest served with sloes: you are the road the curve the arm. ashes will make us rich, coal's the goal. over there, contaminated mountain on fire, catastrophe as error, as world. wear and tear cheaper by the dozen. when lilacs last in the tulips bloomed like hölderlin astronomically, cursorily. onward, christian soldiers, no, stop, do not run east or west, all deities reside within the human breast.

3
in the shadow of snowflakes at the end of the 16th century in the market universal law inside the globe airshafts air columns musical sound of white lime and all together now. cut-phrase hesitation, granite slates, mortars, the lightness about her, her smile in cross section on supporting beams rowing crew frogs in the middle of the house the flour bin eye make-up ochre powder gold powder marjoram ashes of land mammals and sharks because then the waters are troubled and the clay roiled against the wind.

4
a solid body of regular inner and outer structure oh susanna. a crystal in two mirror halves hallelujah how crazy can you get. but her whole behavior and character is hasty, as saussure would put it: feverish. her body of language is a text. her body is a language. because the hummingbird must render blue for blue, blush red, gold. zigzag gliding into february. I'm free, even as lettering with onions and stiff symbols with inscriptions tin signs clay bricks and hummingbirds whereas the many women employed by the company are unpolitical. they unpolitick, they pollunatic, oh polly. in every revolution there is the paradoxical presence of traffic, circulation, blood, military elite, corruption and total destruction. the religion of megamachines demands monstrous sacrifice.

5
a whitish-gray cloud of pigeons in front of a greyish-white pigeon of clouds in january snow, on the other hand the possibility of remembering geometries of absolutely conscious experience. a golden opening at the foot of the column until sunset in season. even the piano fantasy in c minor had two wooden slits. in the market they're at it, lying and screaming. the third kind of embalming is done by washing and rinsing the abdominal cavity without reference to theorems which in turn cannot be validated without time measurement. on march 20th, 1661, at 10 AM, there was an eclipse and, on february 16, a comet could be seen in the constellation of the eagle. the reflection is due to dammed-up smokey emanations from the

second half of the century. petra weaves swift movements, rivets us, lives with us and beats the snow. green berries bloom into red metaphor which stomps, comes to a boil, and isaiah covers the nymph.

ELKE ERB: EIGHT PROSE-POEMS

GRIMM'S FAIRY TALES

A grey sow in a corner. After a long walk in muddy shoes, after an already endless ride in the local, freezing, in a corner of the yard, at the end of the world, why don't you show us your animals, Karl, there she stands in the pale light, peers over the fence posts, a grey sow, in a low pen, at the end of the world.

DIARY ENTRY, JANUARY SEVEN

Visited the jumping Jack, who moves all by himself, and an expectant mother. A fly kept hitting the lamp, buzzing so loudly we could hardly hear what we were saying. There ought to be a Chinese character to put a seal on such adverse conditions and lock them in the bookcase once and for all.

Walls, if you let them, drift down the mountain side and out over the sea. The names of my mountains in Berlin-Center are: stomach ache, belly ache, rotten sleep, damage done.

My new apartment must have dropped out of a capitalist's pocket in the 'nineties. This man, of whom I know nothing else, beheaded an egg every morning, and this man's better half wore for jewelry castiron earrings from the royal Prussian foundry.

To keep ferrets as an environment, why not? They would run back and forth on the bare floor, a good idea, albinos with white fur, they'd look up with inflamed eyes. Even my afternoons would then appear white with small pinkish eyes and take on shape. We'd just have to store everything a couple of feet off the floor to avoid damage.

WHO WENT DOWN THE WELL

After the doctors had nearly given up on him and allowed his wife to visit for hours on end, after the neighbors had said: "He looks like an eight-year-old," after he had nevertheless come home one day, in summer, had learned to walk again, to read in his deck chair and had gotten well, he died, although he had even started, in the fall, to work in the fields again, he died in the winter.

OLD KASPAR HAUSER

If he ventured even just two steps outside his house, the old miller would carry an axe in his belt. He was peculiar. For drinking water he walked to the spring, four hundred feet across fields. He was afraid the neighbors' wells were poisoned. Now and then, on a holiday, a neighbor woman would bring him a piece of cake. He never ate it. When, one day, the pieces were discovered in his house, he admitted his suspicions. He was over eighty when they came and put him in some nursing home in town. He ran away, over eighty, walked back home. The neighbor woman heard the car which took him back again. The second time he walked back all the way from Dresden. He had not done any harm, he said, he just wanted to stay in his own house. But there was nobody to do for him, he would have gone to the dogs: a field, an appletree, a pond, a brook, an empty house.

RETURN FROM THE CONCENTRATION AND DEATH ISLANDS

My face this morning, as if of course I'll die twenty years before my time because I worked underground since the age of fourteen. Where I stand I see, out of every mirror, with open or closed eyes, the ships with the tortured coming in. On the third day, brackish smell, a splash of sun, the return.

UNDERBRUSH

Friday we went to the Pankow woods which in their winter nakedness were the very image of sinister chaos, underbrush all the way to the treetops. We walked for over an hour. Snow lay packed between the scraggly bushes, trampled flat on the path. By a tree a bit off the trail, there was a small boy on his knees, in a navy sweat suit, giggling out of control. Another boy at the edge of the path, was aiming a piece of wood, his gun, at him. A hundred feet off, a third boy aiming at a fourth, while a fifth came crashing through the bushes toward them, hollering: "Don't!"

THE SQUABBLES OF CHILDREN

While you would find my two sisters bickering whenever you looked, I, the oldest by two years, was "quiet" and "reasonable." They drew sparks from each other, and their energy exploded the more fiercely because our environment was too busy rebuilding cities and correcting the past to pay attention to such appeals of childish and immature passion. But while the two of them, rolled into one ball of contention, brushed past me, even I, docile pupil of a narrowly rational and logical system of education, tried, brooding, to fight the stony dreariness, to mine with fuses and dynamite the colossal rock of inevitability before me. Surely it must open, nay change into a magic garden of fulfilment once I had thought everything through and found the password. At bottom, so I remained convinced, everything is simple and in harmony. So what's the big difference between me and my sisters? Doesn't it amount to the same thing whether you fight all the time or hang on to one conviction with all your might? Isn't either absolutely childish? Not a difference of character or even temperament. Noisy or quiet! Don't we know how loud silence can be? But even without meanness, the squabbles of children yank at all the roots they vie to sink into their real estate in the world—a process of development with tests and failures, probations and growth, which is by no means sure to encourage the most desirable and happy traits. Today I hear those childhood squabbles as unmistakable, uninterrupted sobbing.

ASSESSMENT

A writer meets a girl from a small town, whom he calls charming without giving details of eyes or voice, a girl who is in love with a well-built married man (father of children), in possession of considerable soul potential (for instance, a feeling for nature), but alas, in an immoral position from the point of view of society because she just cannot bear to be alone, and now, on top of it all, there's gossip about him, the writer, though without cause. For he remains faithful to his utterly unattractive and even repulsive (her social climbing, careerism, modernism) Xanthippe, as his wistful fatalism seems to require. It's amazing how, as soon as probably "normal" and "fullblooded" people arrange their inner life in this kind of expressive dance or novella, it looks as unoriginal, borrowed, poor, helpless, scrawny and lost as, say, a bit of torn curtain in a huge, grimy dump. Don't they know that in the literary stockmarket you only get rich with home-minted coins, that everything else is counterfeit? The lover, by the way, once she has given him up, will manage a heroic death, while she ends up in the hospital, with a scrawny neck, weeping.

PETRA VON MORSTEIN: SEVEN POEMS

FOR C. S. LEWIS

who imagines, I've been told, that hell is
like the vast slums of Oldham (England)

Can you believe.
That I drove
through Oldham today, that
of all things
a truckload of wobbling
sheep carcasses
got in front of my car
the whole time
taking the streets
I had to take.
By train, I've been told,
you can avoid Oldham.

IN THE CASE OF LOBSTERS

There are
2 methods some put
the live lobster
in boiling
water for the best
taste
but
with a microphone
you can hear screams
of pain if
in the case of lobsters

one can speak of such a thing
as pain

Others
for humanitarian reasons
put it in cold
then bring to the boil

ANTHOLOGY POEM

Yesterday I was
given flowers.

The fact that
I see their wilting
only when
I've gone out for a while
inhibits
my decisions.

At night
I don't need
to put out the light.

THING POEM

Moving out
I was given
a vase.

The notebook was bought
on the island
in the one store there.

You found the striped pebble
on the beach at Aber-Bach, in Wales.

With this pencil
I wrote
things nobody liked, not even I.

Please.
Take off these story tags.
I'd really like
a few things with
qualities of their own.

PRESENCE

Somebody talks
about something and
many listen and
I listen probably
every word would be
the same
if my seat were empty or
if some seat which is not
empty were empty and
somebody
sat in my seat
if I were not
present.

In a big airport
I have even less influence.
Even so
I make a difference.

SUPPOSEDLY INSPIRED

I get no answer
as to the names of
the gardens of Palmyra.

Mr. G., the story
"A Visit to Angelmodde"
will probably not get written.

And then
my "Binsey-Green" poem.

About Aberfan. When
names have meaning
like Vietnam.

JUSTICE

I'm always
most surprised
when
after a trip
a plane trip especially
I meet
in the airport
for example
at a movie
or windowshopping
someone I know

Now I've decided
to be just as
surprised
over the just as

fortuitous
encounters with strangers
even
if I only
pass them

FRIEDERIKE ROTH: FOUR POEMS

AND HOW

1
Just to sit here like this
and listen to the evening
what? To look out over the hills
and not feel
the low sky
its weight

as if we were at the age
when one's happy
and serene
and lighthearted.

2
it's only proper
a pull into bad dreams
no doubt
and questions questions and asking
and getting thrown out
among the firs

which are
nostalgia traps set for us
in a row like soldiers

back then in those were the days
back then as now:

3
the fake gold rings
the tattered clouds strung up.
You've got to
see through it and figure it out and find
what? It's frightful to be in the woods.

Nature, its ramifications, let alone cities
and how minuscule people
rush about there, busily, well, it leaves us
with sharper eyes

4
cold around the heart.
Passion? Looks hilarious
under the microscope.

Again and again
we find somebody who
on occasion would sure like to make a nice family
or at least some kind of
happy couple, what
a scream.

5
What
would we come to if we
gave in to the craze for happiness.
When we've known all along,
between the champagne and fake horror:
power sits in the shoulder blades.

So let's attack
how do you attack a shoulder blade?

6
Nice slick bone-art
is not what I have
in mind, rather set on celebrating
vague orgies of lightness

with the dear dear people
who live on the other side of these hills.

NATURE POEM

                                        (for Manfred Esser)

1
So here lies a stone
and there.
All the things still to be imagined.
An endless web
if
luck would have it:

2
Things come, things go,
above, a quiet glow.
We've soon said it all.
Talk about one thing
and easily lead to another.
We trade heaven in
for earth
and leave the kings
barefoot.

3
The way we lie in the clover
and look up into the sky

as if it were harmless.
Old things won't dazzle us.
You know we've
eaten this mushroom
and can fly,
brazen, without rhyme
or reason. Let's,
for a moment, to the sea:

4
You go on with the good thoughts,
and I'll take care of the narrative.
Where is the pleasure, in sentences
or love?
And why already again
the salt water in your eyes?

5
Our favorite study:
an impossible silliness—
the one with the leaden soul.
I have let all
the threads drop.

6
To climb down
into a flat valley
which at least green
is very dangerous.
Loathsome crawling roots
and screams from a distance:
dirty games
and messed up the sheets.
Already the sunset.

7
The sky falls back.
I'd have done anything for it.
But nature
is always the same.
The game
(a game of tag)
looks good, masterful:
how it leads to the most curious scenes.

8
It must be our pulse rate.
within weeks
centuries rush through us
and rip
the drums off our ears.

9
The old road, the old bed.
We roar with laughter
and forbid mother nature
with all her mushrooms and stones
and her will-take-care-of-it
to invent us.

10
We don't want to be imagined
for anything in the world.
We'd rather think up nature:
Come, my beauty,
the king wants to dance
with you, barefoot through town.
Which I just built.

CHILDREN'S POEM 3

Rather
if I'd found out
ever
I'd tell you
not just for example.

Let me think:
Yes, the mailman, the garbage man.
Yes, there are
trades that make strong poems.

You can write how callused hands
push a strand of hair
out of a face as long as
it's on paper.

My hands
are OK, look
why I take your head
and turn it toward the sun

just when it isn't
falling through the branches

POET'S AUTUMN

1
This early fall,
how it strains, gone mad it seems,
to strangle the
last sunny days.
Only the copper beech
may take its time.

2
Quick nights.
Spanking new adventurers
with the old familiar stories.

We wallow in phrases
sing earth and the bugs.
Why not.
The oblique?
Mygod fear comes
spflashing down on us too.

The baby moans
the wind blows
wine shows no consideration
and later there's always disaster.
Long letter
a chair left empty.

3
All through the summer the poet
arranges to be driven to valleys full of hawthorn.
Behind rolled-up windows
he sadly looks
at the blossoms.
At night he writes
sentences smelling of hawthorn
(as the critics proclaim later).

4
I mean the places in the shade
where snow stays on into June.
Extreme situations.

But we don't turn around
rather risk all in one throw
and laugh

at a little stone and the question
if it's really possible
this out and out useless thing.

In going down
we at least tear open
furrows
and slit the soft earth.

5
How deep water
moves slowly
and ferns
smell so strong from growing on rot.

6
Who can hold on to
the five corners of the blanket?
Who do they think they are
flattering us out of our hopes
blathering about paying our dues
to us
the writers of a theater of crotchety obsessions.

7
Clouds come to mind
and two, three clear brooks
and these spots
which could, what do I know,
represent anything.
I'm not close enough.

8
Everyday
the long-drawn-out terror

the bad mornings
brooding over bits of thought
and evenings and late into the night
pieces, so poetical,
of comedy.

9
On these urgent errands
our heart
has long run off with our head.

If at all
we may as well crush our own skull
and insist loudly, mockingly:
it's a fine thing to do something for Nothing.

As if we were
at another end of the world.

10
Early fall
closes a door.

NOTES ON THE POETS

HEINER BASTIAN was born 1942 in Rantau on the Baltic Sea. After extended stays in Africa and San Francisco, he now lives in Berlin. His important books of poetry are *beobachtungen im luftmeer* (1968), *die bilder sind vor allem nur wie du das rot empfindest* (1970), and *tod im leben, gedicht für joseph beuys* (1972), all published by Hanser Verlag. He also writes art criticism. He has said he wants "images whose cool formal logic points to the laws of language and allows us to discover the grammar of occurrences," but also that images occur as images only in the brief moments when the emotion is stronger than what we are able to say.

48 FIVE GERMAN POETS

CHRIS BEZZEL was born 1937 in Wetzhausen (Franconia). He was living in London when his first book, *Grundrisse,* was published by Luchterhand in 1968 with a preface by Helmut Heissenbüttel, among whose circle of experimental poets Bezzel is usually counted. Other books are *Karin* (Anabas, 1971), *Kerbtierfresser* (Luchterhand, 1972), *Die Freude Kafkas beim Bügeln/Die Freude Mozarts beim Kegeln/Die Freude Bismarcks beim Stricken* (Hanser, 1972), *Salz und Sonne* (Verlag neuer Phantasten, 1983), *weissverlassen steinig* (postskriptum, 1983), and *99 gedichte* (postskriptum, 1987). He teaches linguistics at the University of Hannover and has also edited a *Kafka-Chronik* (dtv, 1983). Heissenbüttel has stressed Bezzel's "alienation of metaphor which appears only as reminiscence" and his "risky attempt to supplant symbolic speech by structural models."

ELKE ERB, born 1938, has been living in the German Democratic Republic since 1949. Two books by her have appeared in the West: *Einer schreit: Nicht!* (Wagenbach, 1976) and *Trost* (DVA, 1984). Her GDR publications include *Gutachten* (1975), *Geschichten und Gedichte* (1976), and *Der Faden der Geduld* (1978). She translates from the Russian and works with a children's theater in East Berlin. She is considered a "difficult" poet in the GDR. Christa Wolf has called her work "insistent, stubborn, authentic" and has praised her "dialectic of nearness and distance, in which close attention does not crowd the object, but allows it to speak for itself." Elke Erb herself speaks of her "eye fixed on the molecule . . . because truth is always concrete and in a working tension to logic and formula. But like the latter, it requires strict precision and clarity."

PETRA VON MORSTEIN was born in 1941 in Potsdam. She teaches philosophy at the University of Calgary and has translated Wittgenstein's *Blue and Brown Notebooks* into German. The poems translated are taken from *An alle* (S. Fischer, 1969). Recent work can be found in *Kürbiskern, Die Zeit,* etc. Not surprisingly, she has called her poems "speech acts" and "instructions in the use of words."

FRIEDERIKE ROTH was born in 1948. She is a most versatile writer. Her books include poetry (*Tollkirschenhochzeit,* 1978; *Schieres Glück,* 1980; *Das Buch des Lebens: ein Plagiat,* 1983), stories (*minimalerzählungen,* 1970; *Ordnungträume,* 1979), and many plays (*Krötenbrunnen,* 1984; *Die einzige Geschichte,* 1985; *Das Ganze ein Stück,* 1986). She has received a number of prizes and fellowships, including the coveted Ingeborg-Bachmann-Preis. She lives in Stuttgart.

EIGHT POEMS

JORGE VALLS

Translated from the Spanish by Louis Bourne

TRANSLATOR'S NOTE. *In a recent poem, the Cuban poet Jorge Valls declares, "Your gut has always been ripped open/ For the world to break into verse." His words aptly describe two aspects of his poetry: the intrinsic anatomical violence of his language and the transformation of the real, even the grimness of a prison cell, into the heightened experience of his own lyric symbology. For when human nature is depraved, Valls translates it into the degradation of the natural world itself. His experience of more than twenty years in Castro's prisons as a dissident for freedom of speech and assembly becomes an exercise in the deepest kind of spiritual survival. The mystery of suffering is often countered by the longing of an almost mystical devotion. The word is extracted, even at the cost of blood, to preserve the memory of love, redeem the pathos of pain, or trust in the future of an adopted son.*

Jorge Valls Arango was born in Marianao (near Havana) on February 2, 1933, the son of a middle-class Cuban mother and Catalan father. He studied liberal arts at the University of Havana but interrupted his studies on March 10, 1952, to join the university protest against the military takeover of Fulgencio Batista. One of his first publications was an article (1952) on this event for the Guatemalan journal Wanima Winak. *He spent most of 1954 in exile in Mexico, worked as a journalist and translator of French and English, and published a poem in the Mexican journal* Humanismo. *In 1955 he became a founding member of the Revolutionary Council, an organization whose ideals, writes Valls, were "democratic, Christian, and socialist, though not Marxist" and which fought against Batista's regime, though he did not join the guerrilla movement in the hills, partially because he found their lawyer leader, Fidel Castro, untrustworthy. He returned from another year's exile in Mexico in January 1959 when Castro's victory was clear, though he doubted the new revolutionary government's attitude toward human rights. Five years later Valls was personally interrogated by Castro for trying to testify for a friend he believed wrongly charged as being a Batista police informer. The friend was shot, Valls imprisoned.*

The human rights organization Americas Watch published Valls' account of his long ordeal: Twenty Years & Forty Days: Life in a Cuban Prison *(1986). The noise of executions and the scourge of beatings and malnutrition were countered by the prisoners' republic and a literature smuggled out of prison. Though much of his poetry was intercepted and destroyed by the authorities, Cristina Cabezas devoted herself to an international campaign to win Valls' freedom, and his poetry won five international awards, including the 1983 Grand Prix of the International Poetry Festival in Rotterdam. Valls was freed in 1984, forty days after completing his twenty-year sentence, and now lives mostly in New York. His books of poetry are:* Donde estoy no hay luz y está enrejado *(Where I Am There Is No Light and It Is Barred: Madrid, 1981); a two-part volume,* A la paloma nocturna desde mis soledades *(To the Night Dove from My Solitudes) and* Hojarasca y otros poemas *(Dead Leaves and Other Poems: Miami, 1984); an anthology of the first book with the same title (Madrid, 1984), with French translations by B. d'Astorg, J. Blot, S. Fosquelle, P. de Laurière, and F. Verhesen (edited by Verhesen) and English translations by James E. Maraniss and Emilio E. Labrada (edited by L. Bourne), plus introductions by Stephen Spender and René Tavernier, supported by International P.E.N. as well as its American, English, and French Centers; and a volume completed with the help of a Cintas Fellowship,* Coloquío del azogamiento *(Colloquy of Quicksilvering: Miami, 1989), a kind of dialogue between mother or father and son in which logic hides and unreal images and environs make up a theater of the imagination. Apart from two unpublished works,*

Encuentros (Encounters) *and* Viaje al país de los elefantes (Trip to the Land of Elephants), *Valls has written several plays, one published*—Los perros jíbaros (The Stray Dogs, 1984)—*a prison novel, an account of his experience of the Revolution,* Mi enemigo, mi amigo (My Enemy, My Friend), *and articles on Marianism in* France Catholique.

In a note to his latest book, Valls gives an organic explanation for his lyrics: "I do not make verses. I shed them in my passing. They are the only honest autography of my life." Apart from earlier twentieth-century Cuban poets such as Emilio Ballagas, the Spanish woman poet Angela Figuera Aymerich, and postwar Catalan poets, especially Pere Quart, Valls' poetic affinities are perhaps most with the Spanish Golden Age, the devotional poetry of Fray Luis de León and the mystics Saint John of the Cross and Saint Teresa. Writing of Valls' first book, Stephen Spender observes: "Sometimes it almost seems that the suffering of our time is the reality which forms the most important subject matter of certain poets. . . ." Indeed Valls portrays the creatures of nightmare, but the zoo of persecution is increasingly mitigated in the later books by the delicate imagery of the doves, deer, lilies, and dew of mystic verse, even if the poet sees new prisons in the alienating metropolis and its lonely cliffs.

I COME FROM THE SQUARE OF FALLEN STARS

I come from the square of fallen stars
Where the houses of the dead look on.
Broken moons gleam in their gravestones
And flags made of rag
Beckon to fairs without stalls.
I cross the borders without touching them,
Their ample patch of shadows,
Hollowing out with words
My long, flayed song.
Above, the fireflies are humming
In crippled and clouded routes.

The last drops fall
Bone by bone on my back.
A depraved thirst makes my eyes
Gape at the faces.
And, meekly, I set out
To return from nowhere.

>   (Ah, my last bowels,
>   My muscles, my lymph nodes,
>   The merciless anxiety of my nails.
>
>   Ah, the quartz lips
>   And the clay teeth,
>   And the magnet at the core of my skull.)

BURN ME

Burn me.
Cauterize the pits of my eyes
With smoking firebrands.
Twist my nerve ends
And beat them with iron hammers,
So they won't leap up,
So they won't cry out.
Force me, master and subdue me.
Thrust my rabid snout
Into the puddles of piss,
And torture my neck till it groans.
Squeeze the brains and the guts
Till they weep,
Till they run, desperate, dissolved,
Now forever overcome,
Last routed rebellion.
Break my posthumous fangs.
Grind my claws.
Liquefy my dust.

Ay! I go forward with coals,
And scratch myself with silex hooks.
When witches are no longer left in my night,
Nor raging scorpions in my belly,
Nor piranhas of sulfur in my looks;
When I am no longer anything,
Not even dung,
Not even reblackened soot,
You will blow with your breath and I shall become air again.
You will gaze in my waters
And I shall be a sturdy orange tree, immaculate
In your regal twilights.

MOTHER, WHAT ARE PEOPLE DOING

Mother, what are people doing
Among steel and stone?
The cyclones slowly pass
Always over their heads
While time grows moldy,
Turning, turning.
I no longer know how many cyclones
Howled by my ear,
Nor through what hole in the ceiling
Death came wailing in
(Death that lay down
Without sleeping on the makeshift bed
Like a sly she-wolf
With her straight, pointed tooth.)
The winter drizzles
Spat in my face
Without a windowpane to cover
My naked poverty.
Where the trees stand tall
People are strolling
While an insect travels

Over my dead clock.
Drilled with silences,
My tongue slowly sweeps
From door to window,
From window to door.
The sides are forbidden;
Now only height is left.
I am gradually sinking below.
They are calling me from above.
Mother, what are people doing
Between the crack and the wall?
Meanwhile, the big rings under eyes
Beat against the rawness of the wind.

WHEN MEN FINISH THROWING AWAY GOD

When men finish throwing away God,
The animals will retrieve Him;
The doves: they have always had Him.
The leaves of the trees
Will be more than happy
For they are going to have Him completely,
Playing in the treetops
Like a child swimming in the foam.
Mother Earth will start to cry
In her long and sweetest rivers,
Because it will be like that day
When He gave His masculine heat
And she felt transfixed and fertilized,
And guessed what the blood could be.
With Him, the stars, like floods,
Will overrun the plains of heaven.
All this if they finish throwing Him away
(They already have Him tamed
And set apart in a little pen.)
But who knows . . . ?

There is a conspiracy between children and flowers,
Between beggars and showers,
Between twilight madmen and the moon,
A whole inferior race of mendicants
Is involved,
And they lift up their rags and sing.
Who knows? They may make a pact with the honeysuckle,
And take Him
Between runners of white lilies to the forest
To be crowned.

WHEN THEY PUT HIM IN THE NARROW ROOM

When they put him in the narrow room,
The last door closed.
Hundreds had closed before:
The door of the light
In the eyes of the swallows,
The door of the water
In the cornlike skin of the beach,
The door of the song
In the wires of the electric lines,
The door of the flower
In the pins of the drilling machines.
Even the door of silence
Was brutally slammed shut
In his face by the loudspeakers.
When they put him in the narrow room,
He heard the last hinges,
Saw the final shadows,
Felt his legs asleep
(But there was no reason to walk any longer;
Truly, he had arrived).
He spoke to himself.
Someone heard.
Then, in a flash, they opened the skylight.

MASSACRE

"What is this bruised moss on the earth?
And that banging of boards in the room back there?
And this spasm of steel in the stomach?
And these iron fingers on the ribs
Begging for mercy
With a voice distorted with groans?
What is this chill invading my arms,
Grabbing me by the shoulders,
Coiling like a little snake round my neck?
What is this sledgehammer
Beating my temples,
And that trail of open mouths,
And of broken teeth,
Of holes reeking with gunpowder,
Of chewed dahlias,
And those quivering owls
Hanging from their own deaths
Unable to shut their eyes?
And those eyes open, always open,
Now forever, always open,
Skewered on the tips of knives,
Floating like jelly monsters
In the liquid of the mire?
What are these long steel needles
Running through a brain?"

"This is one man killed by another.
This is Man,
After they have carried off to the dung heap
All that is left of a man."

RAT

She came from the dung,
Climbing up a stream of urine;
Her face glossy and wet,
Her eyes terrifyingly vile.
She came from the latrine pipe,
Ran diabolically from the deaths
That lived in the stick and the bowels.
A pitiful spattering
Offended my legs.
Then a shock made my flesh cringe.
She jumped up and fled, the tail long and bald,
The mustache disgusting,
Sticky.
I did not want to kill her for she was alive.
She was my sister,
The one who looks most like me,
My sister rat,
Who, in one leap, was lost
In the open belly of the gutter.

WILL THESE THINGS WE DO

Will these things we do
Be any good when we come back from everything,
Even from a sigh;
When the hand returns
From the caress not given?

I love the thorn that infects the finger
For it was born from the stem of the rose.

I love even this misery of mine
So in danger of slipping away
To the foam where the fish cry out.

I love this dust
Winnowing in my fingers
As though it would spread on the face of the wind.
I love the water that falls
And the water that stays,
The humble gray water of washings
(The streaked shadow of wallflowers,
Where the slightest murmur makes a song).

Beneath the herons' wings
The disembodied voices fall.

A bird with a bruised beak
Carves the bleeding ruby of my flesh.

THREE STORIES

JOYCE CAROL OATES

THE WIG

It was in a curious motorized vehicle—resembling a small open-cockpit airplane, or a canoe with wheels—that they drove themselves to Building E in which their friend H. was convalescing.

The hospital grounds were enormous!—far larger inside than one would have judged from the outside. Shading their eyes against the sun they could scarcely make out the fifteen-foot spiked iron fence that surrounded the property.

In the entry to Building E a young black attendant seated at a desk asked to see their passes. His white uniform was stained at the cuffs. He was lean and wiry, with round schoolboy spectacles. For a nervous moment it looked as if he would not allow them into the building though their passes were initialed and seemingly in good order. Then he waved them on down the corridor and returned to his immense paperback medical textbook.

They walked quickly along the lengthy deserted corridor, speaking in whispers. Where was E-18? The numbers on the doors were faint and blurred. They passed a room through whose open doorway they saw an oldish man with gray hair seated at a small table, typing; they passed a room in which a woman of youthful middle age was seated making up her face by way of a hand mirror; another room in which a dark-haired young person—male or female, it was difficult to tell—glanced up hopefully at their approach. They decided that E–18 must have been the room belonging to the man at the typewriter though that man had not appeared to resemble their friend.

And so it was!

H. had changed so drastically, they might not have recognized him. He had aged, he looked bloated, his face had turned gray and puffy like something decomposing in water. Yet his eyes were brightly alert, and his hair though entirely gray now was bushy, springy, almost defiantly full—the hair of a man in the prime of life and in perfect health.

Hello hello hel*lo!*

H. was overjoyed to see them: greeting them warmly, hugging them each in turn, in his old effusive emotional way.

If he noted their expressions of shock he was gentlemanly enough to give no sign.

To their feeble questions of how are you he said, rubbing his hands briskly together, Fine! As you can see!

H. was deeply immersed, he said, in work. Revising a story. A new story? they asked eagerly. No, not new, H. said, in fact not new at all, one of the much-anthologized stories he'd written in a burst of manic energy at the age of twenty-four and ever afterward regretted not having polished. He was adding several new passages, new lines of dialogue. Please read us what you've written, they begged, but at once H. became slyly evasive. He picked up from the clutter of things on his table a sample dust jacket for his next book—sassy lemon-yellow with bold black print—and asked did they like it? Yes they said enthusiastically, they liked it very much, but H. persisted, *Do* you? *Do* you? holding the jacket at arm's length as if to get it into better focus—in the window's light, the glossy yellow surface appeared to catch fire—and they repeated, Yes, though faltering a bit like men caught in lies. H. tossed the jacket down without comment but spoke for several boastful minutes of his publisher's plan for the book: H. was to travel extensively around the country promoting it which he'd never done before but was now really looking forward to. For, after all, things were different now.

They did not ask H. in what way things were different now.

H. changed the subject abruptly. With a sudden wink that screwed up half his face like crumpled paper he said, I know why you boys are here!

They laughed uncertainly but did not ask H. why he thought they were there.

H.'s room was large and sparely furnished: a standard-issue hospital bed, sheets rumpled and pillows awry; grimy white walls and sticky, well-worn linoleum floor; nondescript vinyl-covered furniture. On the dusty window sill were numerous gifts—books, pots of wilted flowers, unopened boxes of candy. On the aluminum typing table were untidy piles of manuscripts and a battered old manual typewriter and, on a hospital tray, a glass pitcher one-third filled with fruit juice and several small paper cups that had the look of having been used. Belatedly H. invited his visitors to sit and asked would they like a drink?—pouring fruit juice into the cups and handing them over before either could decline. Thank you, they murmured awkwardly, but when neither could bring himself to sip from his cup H. seized the wrist of the man nearest him, leaned over, and spat into the liquid. Now you know why you don't want any of my grapefruit juice! he said laughingly.

They all laughed, relieved. It was the old H. after all.

Visitors to Building E were allowed only forty-five minutes. The remainder of the time passed pleasantly if rather slowly: H. insisted upon hearing news, gossip—who of their old friends was where and with whom and for how long and what had become of and did they know anything further of—but paid little attention to details, chuckling, and murmuring from time to time with an air of deep satisfaction, Yes! Of course! When it was time for his visitors to leave they rose shakily to their feet, their limbs strangely heavy, their eyeballs seared in their sockets as if they'd been staring into a blinding light. H. heaved himself up from his chair in a well-practiced maneuver—on the whole, H. was shorter than they recalled but heavier, almost massive about the shoulders and torso—and locked them in bearlike hugs, each in turn. How strong he was, still! So close, the odor of disinfectant made their eyes water; his bristling gray hair scratched like wires against their cheeks.

H. wore an attractive quilted robe and bedroom slippers; flashes of his white bare legs suggested that he no longer had ankles, and that his legs were hairless as a child's. Yet he seemed, for the moment at least, jolly enough.

Wagging his finger he said, I know when I'll see you boys again!

They did not ask when he thought he would see them again.

At the end of the long empty corridor they paused to look back and sure enough their friend stood swaying in the doorway of his room, grinning and waving.

No nurses or attendants were in sight. The young black man in the entry had disappeared. In their agitation they couldn't remember how to operate the motorized vehicle and after several vexed, frustrating minutes they decided, despite their exhaustion, and the considerable distance to the front gate, that they would walk.

BIOPSY

In her girlhood of some years ago she'd been a serious swimmer, so all this constituted really was swimming in uncharted unrefereed waters, the space of time she would have to endure between what her doctor took pains to call the "surgical procedure" (*not* an "operation") on Tuesday morning and the ringing of her telephone Thursday morning, 8:30 A.M. approximately, when his office would receive the laboratory test results from Philadelphia. So it wasn't infinity but merely forty-eight hours more or less of which she might spend as many as fourteen hours unconscious, two nights of seven hours sleep each (if not natural sleep then drugged sleep— she would be taking a morphine derivative for the pain), which left thirty-four hours of consciousness of which she could count on as many as ten hours spent in the preparation and consumption of meals and the kitchen cleanup afterward if she stretched things (why not telephone people as soon as she got home, invite them for an impromptu party tomorrow evening, spend much of the afternoon tomorrow preparing an ambitious meal then the evening would be entirely taken up in eating, drinking, sociability, no one would guess the state of her nerves and their not knowing might cancel out that state—swimming isn't always against the current after all), which left twenty-four hours, a nightmare of time if you stared unblinkingly into it, but she had her work, and she had the tasks of her household, and she had reading, and she had the telephone—she might call, for instance, her mother (with whom since her father's death she was close, yet distant—close in her feeling for her, distant in her guardedness about revealing much of herself to her), and there were friends of course, and former lovers if it came to that, that degree of desperation, but she was resolved that it would not. For there was, again, her work, her work that *was* her in the deepest and most abiding sense, and even if in the middle of the night she found herself incapable of sleep she could force herself to work—had she not done so upon other difficult occasions? And should that strategy fail she would leave the house like stepping boldly out of her skull and walk (she was walking now in fact) and walk and walk until her lean muscles ached with a pleasant tiredness and her body was bathed in a dreamy haze of perspiration and sleep would have become unnecessary.

Say she has gotten through the nights. Say she has gotten through forty-seven and one-half hours. Thursday morning and she has been awake since dawn showering calmly and grooming herself in that matter-of-fact way she has cultivated over twenty-odd years to accommodate, not loneliness, but aloneness, the primary fact of her life, and now she is staring out the window warming her hands around a mug of hot tea reasoning that it is only a matter of duration now, a final effort, the shore at last visible beyond the choppy waves, and the land beyond it, so rich, suddenly, with possibility—how like tracery it is, that shore she has never glimpsed before, that land, that horizon, as if an idea only, her idea, waiting to be filled in.

BAD NEWS

Bad news, and at a bad time. So she resolved not to think of it now. Not now.

Bitterly she reasoned: it isn't fair. Not now.

So she prepared the apartment in readiness for him, her pulse fast and jaws clenched for him!—for him!—all for him!—in her bitter resolve cleaning, vacuuming, polishing the kitchen floor until the old tiles shone—for him!—the fresh-picked wild daisies on the little dining room table in the little dining room alcove and the hardcover books, magazines, professional journals casually arranged on the coffee table and the wine chilling in the refrigerator and the salmon salad and the expensive greens and the chocolate mousse and the fresh-ground coffee for him!—and in the bedroom that cramped sunless little room the floral quilt and the floral curtains and the floral sheets smelling of newness for him!—all for him!—and in the bath preparing herself she hears the telephone ring and her heart contracts with the knowledge that she must answer it but will not answer it, not now, not tonight, she must answer it but will not answer it: not bad news, not at such a bad time. They have no right she's thinking angrily, they have no claim she's thinking as she runs the faucet hard so that she can't hear the telephone ringing and in any case she will fly home for the funeral, there will be tears, and grief, and commiseration, but not tonight.

So she takes the receiver off the hook, and it's off the hook for the entire evening, and the night, and she's happy as he likes her, and beautiful as he likes her, and the evening goes perfectly, or nearly, and the night, her lover in the night, as in her fiercest dreams she'd hoped. And flying westward to home the following day she's thinking yes, he loves me, he must love me—though as it turns out he won't marry her anyway, in the end.

TEN POEMS

CRAIG RAINE

A MARTIAN SENDS A POSTCARD HOME

Caxtons are mechanical birds with many wings
and some are treasured for their markings—

they cause the eyes to melt
or the body to shriek without pain.

I have never seen one fly, but
sometimes they perch on the hand.

Mist is when the sky is tired of flight
and rests its soft machine on ground:

then the world is dim and bookish
like engravings under tissue paper.

Rain is when the earth is television.
It has the property of making colours darker.

Model T is a room with the lock inside—
a key is turned to free the world

for movement, so quick there is a film
to watch for anything missed.

But time is tied to the wrist
or kept in a box, ticking with impatience.

In homes, a haunted apparatus sleeps,
that snores when you pick it up.

If the ghost cries, they carry it
to their lips and soothe it to sleep

with sounds. And yet, they wake it up
deliberately, by tickling with a finger.

Only the young are allowed to suffer
openly. Adults go to a punishment room

with water but nothing to eat.
They lock the door and suffer the noises

alone. No one is exempt
and everyone's pain has a different smell.

At night, when all the colours die,
they hide in pairs

and read about themselves—
in colour, with their eyelids shut.

IN THE KALAHARI DESERT

The sun rose like a tarnished
looking-glass to catch the sun

and flash His hot message
at the missionaries below—

Isabella and the Rev. Roger Price,
and the Helmores with a broken axle

left, two days behind, at Fever Ponds.
The wilderness was full of home:

a glinting beetle on its back
struggled like an orchestra

with Beethoven. The Hallé,
Isabella thought and hummed.

Makololo, their Zulu guide,
puzzled out the Bible, replacing

words he didn't know with Manchester.
Spikenard, alabaster, Leviticus,

were Manchester and Manchester.
His head reminded Mrs Price

of her old pomander stuck with cloves,
forgotten in some pungent tallboy.

The dogs drank under the wagon
with a far away clip-clopping sound,

and Roger spat into the fire,
leaned back and watched his phlegm

like a Welsh rarebit
bubbling on the brands . . .

When Baby died, they sewed her
in a scrap of carpet and prayed,

with milk still darkening
Isabella's grubby button-through.

Makololo was sick next day
and still the Helmores didn't come.

The outspanned oxen moved away
at night in search of water,

were caught and goaded on
to Matabele water-hole—

nothing but a dark stain on the sand.
Makololo drank vinegar and died.

Back they turned for Fever Ponds
and found the Helmores on the way . . .

Until they got within a hundred yards,
the vultures bobbed and trampolined

around the bodies, then swirled
a mile above their heads

like scalded tea leaves.
The Prices buried everything—

all the tattered clothes and flesh,
Mrs Helmore's bright chains of hair,

were wrapped in bits of calico
then given to the sliding sand.

"In the beginning was the Word"—
Roger read from Helmore's Bible

found open at St John.
Isabella moved her lips,

"The Word was Manchester."
Shhh, shhh, the shovel said. Shhh . . .

IN THE MORTUARY

Like soft cheeses they bulge
sideways on the marble slabs,

helpless, waiting to be washed.
Cotton wool clings in wisps

to the orderly's tongs,
its creaking purpose done . . .

He calls the woman "Missus,"
an abacus of perspiration

on his brow, despite the cold.
And she is the usual woman—

two terra cotta nipples
like patches from a cycle kit,

puzzled knees, finely
crumpled skin around the eyes,

and her stomach like a watermark
held up to the light.

Distinguishing marks: none.
Colour of eyes: closed.

Somewhere, inside an envelope
inside a drawer, her spectacles . . .

Somewhere else, not here, someone
knows her hair is parted wrongly

and cares about these cobwebs
in the corners of her body.

LAYING A LAWN

for Ian McEwan

I carry these crumbling tomes
two at a time from the stack

and lay them open on the ground.
Bound with earth to last,

they're like the wordless books
my daughter lugs about unread

or tramples underfoot. I stamp
the simple text of grass

with woodwormed brogues
while my daughter looks on,

holding her hard, muscular doll
by its only leg . . .

She hands me a caterpillar
rucked like a curtain, just as

one day she'll bring me teeth—
segments of sweet corn

in the palm of her hand.
While she faces me, I needn't see

the thin charcoal crucifix
her legs and buttocks make,

only the hair on her body
like tiny scratches in gold,

her little cunt's neat buttonhole,
and the navel's wrinkled pip . . .

For the moment, our bodies
are immortal in their ignorance—

neither one of us can read
this Domesday Book.

THE TRAIN SET

"The SS went to a great deal of trouble on the public relations side: there was background music, lorries ready to transport women, children and the infirm, while prisoners were at hand to act as porters; in short an atmosphere of relative welcome greeted the Jews when they got out of the cattle trucks . . ."
—Lucien Steinberg: *Jews Against Hitler*

A broken buffer rolls
in the palm of my hand,
like a collar stud
retrieved from history.

I take out two bridges,
goose-pimpled with rivets,
and notice that a child
has cleverly blackened

the underside of one
with smoke from a candle.
There is a trunk, labelled
WINDSOR, ENGLAND, EXPRESS,

but the tracks were made
in Germany. Fit them
together how you will,
they form a pretzel . . .

Better just to leave them
in confusion, in the sad
geometry of flight.
They will never recede

like a xylophone.
On the little trolley
for *Bücher & Zeitungen*,
the papers are in scale:

De Telegraaf's date
is 4th of Feb, 1929.
The fine print bulges
under the dealer's glass,

as if I were looking
through unbidden tears.
Die Woche advertises
PARKER Duofold Pens

below the photograph
of a skiing girl
who tucks away her sticks
like a sergeant major.

The signal still salutes
as smartly as ever,
with the reds and greens
of a medal ribbon.

All the rolling stock
is in mint condition,
perfect on barbell wheels:
Nichtraucher, Schlafwagen,

Raucher, Deutsches Benzin.
I fiddle with the couplings
like a father dressing
his child in a hurry,

the spinal column there,
alive, under my hands.
I'm suddenly touching
a boy in pyjamas,

the ghost in the machine.
Here is the key that makes
those pistons shadow-box
so fast and uselessly.

RICH

She owns that thoroughbred
smouldering under his blanket,
those binoculars *en croûte*
kept warm by her breasts,

and these are her eggshells
cracked on the kitchen table
like an umpire's snail
of cricketers' caps.

I woo her with words,
I number her many possessions:
seven rings by the soap,
two on top of the Steinway,

her grinning grand
with its dangerous fin.
And the boy belongs to her
who drives his toast

round a difficult bend
with multiplied lips,
or settles to sleep
by sipping his thumb.

She cannot ever know
the extent of her riches,
the mother of this naked girl
who stretches sideways,

a tired toxophilite
bending her invisible bow.
Custodian, I guard
and garner up each treasure.

She is the giver of gifts
without number or name:
this flock of crumbs
on her tablecloth,

the goldfish mouthing
its mantra, the pungency
of pencil sharpenings,
perfume by Paco Rabanne.

Her cattle are children,
each with a streaming cold,
and this is her bull
drooling over his dummy,

his angular buttocks
crusted with cradle cap.
She gives them grass.
They drink from her bath,

or gravitate towards the pond,
her snooker table, torn,
where only one player
attends to a solitary red.

And there, beyond the books
on the windowsill, her floods,
transforming the world
like the eye in love.

I am the steward
of her untold wealth,
keeper of the dictionary,
treasurer of valuables,

accountant and teller,
and I woo her with words
against the day of divorce.
Mistress of death and of birth,

she owns the trout
tortured with asthma
and this field in spring,
threadbare with green.

IN MODERN DRESS

A pair of blackbirds
warring in the roses,
one or two poppies

losing their heads,
the trampled lawn
a battlefield of dolls.

Branch by pruned branch,
a child has climbed
the family tree

to queen it over us:
we groundlings search
the flowering cherry

till we find her face,
its pale prerogative
to rule our hearts.

Sir Walter Raleigh
trails his comforter
about the muddy garden,

a full-length Hilliard
in miniature hose
and padded pants.

How rakishly upturned
his fine moustache
of oxtail soup,

foreshadowing, perhaps,
some future time
of altered favour,

stuck in the high chair
like a pillory, features
pelted with food.

So many expeditions
to learn the history
of this little world:

I watch him grub
in the vegetable patch
and ponder the potato

in its natural state
for the very first time,
or found a settlement

of leaves and sticks,
cleverly protected
by a circle of stones.

But where on earth
did he manage to find
that cigarette end?

Rain and wind.
The day disintegrates.
I observe the lengthy

inquisition of a worm
then go indoors to face
a scattered armada

of picture hooks
on the dining room floor,
the remains of a ruff

on my glass of beer,
Sylvia Plath's *Ariel*
drowned in the bath.

Washing hair, I kneel
to supervise a second rinse
and act the courtier:

tiny seed pearls,
tingling into sight,
confer a kind of majesty.

And I am author
of this toga'd tribune
on my aproned lap,

who plays his part
to an audience of two,
repeating my words.

GAUGUIN

They going upstair
take longtime lookit shedownstair.

They going upstair
so hedownstair go plenty upstair.

He stickyout number2tongue,
because he magnetized to she.

Which she hide in shesecrets,
because she magnetized to he.

They making the mirror, shhh,
numberonetongues completely tied.

Shebody making the horse
and the frog, the safe shescissor,

the squat on sheback, showing
shekipper tenminute longtime,

till he cry like a candle
and heflame blow out.

*Handmake Kodak man, come back,
my secrets are sorry with oil.*

PLAIN SONG

There was the chiropodist,
whose wife had a tapeworm
or a fallopian cyst,
and there was my father

reading his tea-leaves.
I was hidden behind the sofa
and could only see a turn-up
and one ox-blood Saxone loafer:

I'd taken all my clothes off
for the lavatory, hours before,
and now Mr Campbell had come.
My mother shut the kitchen door,

firmly, like a good Catholic,
snubbing the two Spiritualists.
I imagined the best china
in my father's massive fists:

a woman in hoops and crinoline,
the dainty rustic handle,
the Typhoo hieroglyphics,
the fate of Mrs Campbell

whom mother felt so sorry for.
My father went into control
and I listened, out of sight,
to the jumbled rigmarole:

someone was passing over
to another world of love.
Why do I only now discover
the woman's thin voice

saying she will be missed,
my father's eyes rolled back
and Mr Campbell's unhappy mouth,
open like a bad ventriloquist?

THE MAN WHO INVENTED PAIN

He lifted the wicker lid
and pigeons poured
past his hands,

a ravel of light
like oxygen
escaping underwater.

Loss of privileges
in peacetime; in war,
a capital offence.

He offered no defence,
simply composed
a non-existent life

in letters home,
enough for a year,
to be posted in order,

of which the last began:
Dear Mother, Dear Dad,
Thanks for yours.

Today, a Tuesday,
we shot a man
at 0800 hours.

Try to imagine,
if you can,
the subdued feel

of a Sunday morning
and the quiet clash
of a dixie lid,

lifting and lapsing
like a censer
at mass.

Imagine held hats,
blown about hair
and the firing squad

down on one knee,
close enough to see
his Adam's apple

genuflect
just once
before they fired.

And then imagine
the rest of the day:
the decent interval

before the men
began to form a queue
with mess tins,

the way in which
the day remained
a Sunday until dark.

Things were touched
with reverence.
Even the sergeant,

feeling for fags
in his battle dress,
patted his pockets

uncertainly,
in turn, and again,
as if he'd forgotten

the sign of the cross
and the captain
on a canvas stool

sat like a priest,
with praying eyes
and inclined head,

while his batman cut
and curls fell
all over his surplice.

Imagine the sun
waking the flies
to a confessional buzz

in the camp latrines,
and each latrine
a taut box kite

waiting for wind
on the kind of day
a man might read

the Sunday paper
by his pigeon cree,
or nervously

walk out to bat
and notice the green
on a fielder's knee.

THREE POEMS

CHARLIE SMITH

IN VIRGINIA

I stalled a while watching a bay horse
crop grass in a small lot
near the highway. Blue spidergrass
and bitterweed flourished among piles
of rusted implements, among a hay rake with tines
like the delicate rib bones of an ancient fish and old lumber.
I thought, the way one can, that the new loss
of your love
might become a permanent sadness,
and I was sad
because I knew it wouldn't. We had not lingered
one evening in Virginia
to watch a bay mare, her coat roughed
already for winter, crop the sparse bahaia grass
in a cluttered lot—so it wasn't memory
that touched my sadness, though
what amplified the loss
was ample enough. And it wasn't
what might have been,
because I couldn't picture you
in that place, where a sly breeze
pretended to steal the smallest leaves, and the horse,
I saw, was old, and lame. We imagined children

and hard work, waking by the lake
in Michigan, but we got
to none of these. I read a passage once
about a form of chastity
that acknowledged,
but abstained from pursuing,
the beauty of the world, and I thought of the afternoon
I turned to you on the beach at Pamet Roads
and saw your face for an instant
shining like all I would ever love
or had ever loved, and though the moment passed
like one of the thin green waves
skittering in, I knew I would give myself away, I knew
I had already given myself away
to what I saw shining in you, that I would make the mistake
I had made before,
and would probably make again, of believing
your face, your voice, your
history, was the thing
itself,
and so wind up lost. They say it isn't the squalor
that kills, it's the beauty. It's what—and all—
on a bright day
when as we watch the light tremble in sheets
on the pond, we surrender to,
and so go dumbly down,
and are ruined,
and return from dazed and chastened,
as a callow boy caught thieving in an orchard
will glance at the bootless blue sky,
at the treasure of ripe fruit
filling his shirt, and descend
into the irate farmer's world,
unable to explain anything—
not his theft, or himself,
or what makes him what he is.

SLAUGHTER STREET

One time years ago a man tried to sell
me a girl he claimed
was his daughter; with his thumbnail
he pushed the lank, loose hair away from her face,
he traced
with his forefinger the proposal
of her lips, the thin lifted
shills of her breasts. I think
I wanted to hit him, but I didn't.
I said yes, and took her away
under the mimosas that were wet with the afternoon's
rain, thinking that like a cat
I couldn't keep, I would set her loose
in the desert. But she wasn't a cat
and from that town it was miles
to the desert—we wound up, somehow,
on a street whose enamel gutters
ran in the mornings with the blood of slaughtered hogs,
drinking at the tin bar
of a dark place, where a three-piece band
played music compounded of gravel
and sarcasm, and women in ripped dresses
danced wildly, as if in torment, with thin men
who cracked pecans
between their teeth as they danced, and laughed
at nothing. I remember for one moment
I looked at her small face
and thought I saw my own romantic notion
of promise fall from it, but maybe
I didn't see anything at all, and maybe she had stopped
caring long before
I blundered by, but whatever I might have thought
I was soon lost to—quickly, quickly—so that
when I looked again the gray dawn
was getting to its feet in the doorway,
and a man I had never seen before snored on top of the bar,

and the wild women
had retreated into the mossy places
they hid in during the day;
and she was gone, without blessing
or any enterprise I could muster, lost
to the claims of the street,
from which came now the squeals
of pigs, as the quick knives
opened the large vein in their throats,
and allowed
what they had been keeping to themselves
to pour forth.

CYCLES

I think it is well known now
how you can take one part of the country
and re-erect it somewhere else, how the abrupt, snow-streaked
mountains of New Mexico hang in a dark corner
of your uncle's hall. And some morning in late winter
the trees are diadems of ice,
the way they were once, miraculously, on the river in Florida,
when you were a child. It is not
that we wish the courage
to ask a co-worker
for the loan of her beach house, it is not
that. Or the mad woman,
slick with terror and grease, interrupting mass—
that something in here
would take hold suddenly, and calm her. O the priests
do their duty
because they are pros. And the world
was always crumbling fast; it is not that we wish the world
other than it is. My friend asks
what's your hurry? we are all headed
to the grave. Only, I guess,

I seek a certain rhythm, the dumb bob
of a pigeon's head, the old friend
calling from Kansas, the muck of love—
its resignation and exuberance—the red flowers
returned to the trees
as if winter was nothing to them, nothing at all.

CATCH ME GO LOOKING

LAURA MARELLO

One

I watched the feet make flat smudged shadows that pushed against the crack at the bottom of the door. Into the parents' room. They are going. The parents' room.

The sirens came for—

I opened the door on the shadows. Dad in his yellow terry-cloth robe standing there, threw me back into bed, grabbed me by the shoulders and steered me, stuffed my bear into the hole in my stomach. Stay there. Stay in that bed.

They wouldn't tell me who the sirens came for.

Gail in her room too, shut in, crying, screaming, sirens, screaming, throwing pillows at the door, threatening to leave out the window, threatening to lock them out, threatening to cry herself to death, threatening to be helpless.

They wouldn't tell me who the sirens came for.

The shadow of the window frame got big, picked up its bushes, Dad's carlights chased it around the wall of my room as he turned the corner. Gone. He's gone. Go looking.

They wouldn't tell me.

Gail is asleep now, pillows on the floor. Lights left up, to ward off burglars, to watch the stillness, to catch me go looking. Looking, go looking.

Pepper wagging his tail, keeping secrets, wanting some water, to be let out into the garage, into the garden, to bury.

They wouldn't tell me who—

My footsteps louder in the silence. Footsteps screaming in the silence. Screaming silence. Sirens screaming, sirens, silence.

They wouldn't tell me who the sirens came for.

Go looking then. Parents' door locked. Keep out, no smoking, closed on Sundays. The back way then. The bathroom. Snails rolling in the sweaty morning grass. Step between snails through the grass, through the garden, to the unlocked window, through the empty space between glass, feet on the tiles, mud on the tiles, down inside the white sunken tub. Safe. Home free. Olly, olly oxen.

They wouldn't—

Mirror says it's empty. No one in the room. Tile slippery, feet wiped clean in the rug of the walk-in closet, in the must of the moth balls, silk-lined suit jackets, hat boxes, picture albums, Dad's machete leaned into the corner pointed down, poles with wooden blocks supporting wing tips, wicker clothes hamper, tissues smudged red with lipstick wadded, in wastebasket.

None of these unfamiliar clues. None of these sirens. Who were the sirens? For who, for the sirens were, for the sirens who. Who were for the sirens, for the who, who were the sirens for?

They wouldn't tell—

The dresser drawers closed, bed unmade, pictures. A woman with her child behind her shoulder. A little girl etched in pajamas, holding her limp doll, her socks coming off. Purple flowers framed on someone's kitchen table. Not our kitchen table.

All of these the same. Except the bed. She always makes it. But they are going back to sleep? It's four, only.

They wouldn't—

Dad's drawers: handkerchiefs, underwear, old coins, cards, sweaters, pajamas, work pants, old wallets, cufflinks. No.

Mom's: bras, nylon underwear, hose, shirts, heating pad, old sheets, slips, big red hot-water bottle. No.

Sprawled out on the bed illegal way, feet on pillows. No. No clues, no sirens, no screaming, no, sweaty grass leaving traces, gone looking, cracking something cracking under them, under pillow, oh no. What have I broken? Only a note, sounding stiff, I slip it out from the slit between bed and pillow, unfolded, read:

DEAR VIC
(Dad's name, Mom's writing, yes, yes.)

I'M SORRY TO HAVE TO TELL YOU. LIKE THIS, I MEAN I CAN'T CONTINUE LIV—
Tell, they wouldn't—

—HOW THE HELL DID YOU GET INTO THIS ROOM!
—Don't hit me Dad! DON'T HIT ME! and I run for the bathroom. But he grabs me around my stomach. I double over, put my hands over my head, DON'T HIT ME! and I'm choking. He grabs the note, lets me loose, rips and rips and his eyes get so big that I remember to run. RUN!—They wouldn't—RUN! Out the door, down the street (sirens screaming) RUN! across the street, down the next, RUN! (they scream) RUN! (screaming sirens) RUN! (sirens) RUN! RUN!

*

sheee yadda yadda
sheee yadda arta

sheee yadda yadda
artack.
yadda artack.

yerma yadda
sheee yadda yadda
yartack
yadda yartack.

yerma yadda yadda
yerma yadda yartack.
sheee yadda yadda
yadda yartack.
yerma sheeeyadda, yadda yartack.

—YOUR MOTHER, SHE, SHE HAD A HEART ATTACK. Corie, are you listening to me?
—Yes, Dad.
SHEEE YADDA YADDA
sheee yadda
YARTACK.
YADDA YARTACK.

YERMA YADDA YADDA
YADDA YARTACK.
YERMA YADDA YARTACK.

—She's dead, Corie.
—Yes, Dad.
—Go to your room now.
—Yes Dad.

Sheee yadda yadda
yerma yadda yadda
sheee yadda yadda

*

—She said I didn't have to.
—Why would she have told you, dear?
shee yadda yadda
—She told me that if she ever died, I didn't have to go see her at the funeral home, and that you were gonna make me go.
 I mean, that if you asked me to go, I could say no. I mean, that I didn't have to go. She said so. She said, she said so.
—What else did she say?
—She wanted her ashes strewn all over the Appalachian Mountains.
—You know how much your mother loved both you girls, Corie.
—I know, Granny.
—She would have done anything for you two girls.
—I know, Granny.
—And your father has such a burden now trying to raise you without any help.
—Yes Granny.
—That you're gonna have to be as good to him as you can.
Yes Jesus.
—And not get into trouble or worry him about boys till you're old enough.
Hallelujah, brothers and sisters.
—And you know your mother was worried about you.
Oh sweet Jesus
—And talking about you as she died, 'cause you're the baby of the family.

Praise the Lord.
—And she knew you'd have the worst of it.
Amen.
—So you be good to your daddy now.
Jesus Christ, Amen.
—And go to the funeral home to be respectful to your mother.
—She said I didn't have to go.
—Be respectful to your mother's memory, Corie.
—She said so.
—Do it for your father, Corie.
She said so.
—For your father Corie, not your mother.
Said so.
—For your father, Corie.
—NO!

*

Mom's blue sneaker with the white laces hits me on the calf. Gail is scared. Karen will not say.

We shut the door fast, all of us, piling up the desk chairs against it, getting ready by the window. We hear a spiked heel, hitting above the doorknob, landing on the carpet, the other flying by into the bathroom, knocked flat against the linoleum.

He is throwing them, father, he is throwing her shoes. Sandals brown and leather, flung down the hall by straps, loafers hurled onto the kitchen tile, bouncing and skidding into the dog's bowl.

He is throwing them, her shoes, my father, he is throwing her shoes.

The beds aren't made. The dishes from after school are in the sink. Make-up is on the counter in the bathroom, toothpaste dabbed in the wash basins. The oven isn't preheated. The TV dinners are not out of the freezer, and there are many more shoes. HER shoes. Flying into the den, into the dining room, knocking up against the chair legs, slippers flapping down the hall. Her walking shoes, her church shoes, her shoes, her shoes, and he is taking the rest with him, father, with him, out to the front lawn, throwing them at her car, rubbing the heels into the dicondra, breaking the fronds of the yucca tree she planted. They are coming fast, all of them, all her shoes, father, they are coming, he is throwing all her shoes, all her shoes.

❋

I can't sit in the first pew. I can't let the minister see my face. See mine, next to Dad's next to—

Sheee yadda yadda.

He would be able to tell then, he would know. Maybe he was the person she (forgive me—) the person she could—

Tell, they wouldn't—

I can't sit there. MINE! In the first pew, with my father, next to my father's. I can't let him see, (forgive me—) MINE! I can't—

—The Lord is my shepherd; I shall not want.

Dad breathes too heavily, too fast. Granny takes her gloves off, reaches for the handkerchiefs in her black patent-leather purse. She will pat on her compact in the lobby afterward. It is better to leave them there, in the front.

—Yea, though I walk through the valley of the shadow of death, I will fear no evil: for thou art with me.

Gail is going to cry too hard, is going to stop crying after this, is never going to cry after this, is never. Sucked into the middle row of pews, her friends on the other side of her, of me, in my meadow dress that Mom sewed. Green, white, flowers, white, white.

(*Well of course, let her wear it, Dear. It's spring isn't it? She would have wanted her to wear that. That was her spring dress.*)

Was. Would have wanted.

—Surely goodness and mercy shall follow me all the days of my life: and I will dwell in the house of the Lord forever.

I wonder now, who Shirley Goodness was. I wonder now why there was such an urgency for dinner at six. Such an urgency to be held. Such an urgency in having everything explained, everything righted. Righted.

—. . . to honor the memory of Rebecca Brigman Barsoc.

Barsoc. Barsochini. Barsoc.

(*Because they're relentless in the army. Relentless? They'd tease you too much if you had a name like that*).
Barsochini.
(*Barsoc is more solid. More American.*)
They would have. If you were to have. Is more.

—. . . absent from the service. It is as she requested. Perhaps she wanted to be treated, always from her death, as a memory.

I remember her sitting at the kitchen counter, hand over her mouth, finger under her nose, staring through the dining room window. Staring through the ivy vines, through the yucca tree, the driveway, through the kids playing sewer tag. I remember her, at the counter, shaking. Her coffee shaking, her cigarette shaking, her hand over her mouth, hiding, shaking, staring through the window. Staring through the other houses like ours, staring through the brown hills, staring through the dryness, through the heat. Staring past all that. I remember her staring past all that.

—. . . the memory of a fine, strong, patient woman, a deep woman, with a great love for her family and her neighbors.

Only those times when I thought she should have stopped a minute. (*Why are you crying Corie? / I don't know. / Well be quiet then, your father and Gail are trying to sleep.*)
Stopped to be silent, stopped to look. She did that most always. She did that too much. Looked too much, stopped too much, stopped short.
(Sirens screaming)
Why are you crying, Corie?
MINE!
Well be quiet then, your father.

—. . . she came to me for counsel, soon before her death. She was an intuitive woman. She sensed a passing through of some great change, of some great deed that was to be done.

Yerma yadda yadda

She tried (forgive me) tried then. Its ours now. MINE! Ours. She has the why of it. I the how. Lie down. There now.

—. . . We will all respect and honor the memory of this woman, and what she has given us. We will continue, as she would have wanted us to do, even though we grieve. Let us pray.

❀

I am not a poor baby. I do not want your fingers pinching my cheek. I am not hungry. I do not want some of your tuna casserole, some of your pumpkin pie, some of your spring vegetable salad, some of your roast that can be refrigerated and used for sandwiches, some of your bu'ndt cake, some of your potato salad, some of your party, some of your sacrifice.

I did not put your potted plants with cedar chips and ceramic rabbits in my room, on the piano, next to the buffet. I will not keep the announcements under my bras in the first drawer. I did not hang the newspaper clipping on the bulletin board. I have not cried. I have not listened to you tell me how brave I am. I will not be brave. I will not grow up to look just like my mother, be just like my mother, die just like my mother.

I cannot watch the dog sit at the window and cry. I cannot introduce myself to the people sitting in the living room, to the woman vacuuming, to the people eating in the kitchen, to the people gossiping in the laundry room, to the woman spraying ammonia on the bathroom mirror.

I will not play guitar for the guests. I will not go shopping for summer clothes with Mrs. Wilson on Saturday. I will not spend the night at Judy's the week at Newport Beach.

I will not look for my mother when I get home from school. I will not cry at night. I will not dream she is alive. I will not remember what she looked like. I will not try to remember. I will not go through the picture albums. I will not watch home movies. I will not try to remember, try to understand, try to explain, try to believe.

Do not stroke my hair, do not pat my shoulder, do not rub my back, do not tell me you are sorry.

❀

NOW I WAS BORN IN TENNESSEE
AND I WAS MARRIED WHEN I WAS THREE

MY HUSBAND COULDN'T WAIT FOR ME
'CAUSE HE WAS FORTY-SEVEN

—Turn the record off, Corie.
—Why?
—Turn it off!

MY KINFOLK UP AND HAD HIM HUNG
SAID I WAS MARRIED FER TOO YOUNG
SAID I'M A SLAVE 'TIL I BECOME
A GROWED UP GAL OF SEVEN

—She's my mom too! I want to hear her sing!

SO I UPPED AND I MARRIED MY BROTHER-IN-LAW
I DIDN'T HAVE TO ASK MY PA
MY MA SAID SHUCKS, HE AINT YOUR PA
YOUR PA'S YOUR UNCLE FUD

—You're gonna get Dad upset. They recorded those records on their honeymoon.
—He said I could listen to it. He can't hear it anyway. He's vacuuming.

NOW, UNCLE FUD SAID NO HE WEREN'T
YER COUSIN LUKE IS THE GUILTY VARMINT
HE WAYLAID MA THAT NIGHT CONSARNIT
GOT RELATIVES IN MY BLOOD

—He's crying, Corie!
—No he's not!

NOW, THAT MAKES MY COUSIN MY SISTER SUE
DARNED IF WE KNOW WHO IS WHO
MY NEPHEW IS MY UNCLE TOO
WE'RE JUST ONE BIG HAPPY FAMILY

—Turn it off!

—But Karen wanted to hear it, Gail.
—She can hear it another time.

NOW I GET TIRED OF MY BROTHER-IN-LAW
SO I THOUGHT I'D TARRY
WITH MY NITWIT UNCLE ON MY NEPHEW'S SIDE
MY NITWIT UNCLE HARRY

—You really think you're my mother now, don't you Gail?
—Turn it off.
—You think you're so big.

HE SITS AROUND AND HE WITTLES WOOD
HE GOES A BLEH BLEH BLEH BPT HE DOES IT GOOD
I'D SURE MARRY HIM IF'N I COULD
BUT HE'S GOT SIX WIVES ALREADY

—Turn it off.
—She's a broken record.
—O.K., that's it.

SO I DON'T KNOW WHAT UNCLE HARRY'S GOT
THAT MAKES THESE WOMEN HANG AROUND A LOT
HE CHASED ME ONCE BUT I GOT CAUGHT
WAY DOWN IN TENNESSEE

—You're only gonna wreck your own record player, Gail, that's dumb. We'll take it to Karen's and listen to it.
—It's not leaving the house.
—Why not?
—It's irreplaceble.

WELL THAT'S JUST THE REASON I LONG TO BE
IN THE HILLS OF TENNESSEE
WITH A BOOWITHA BOOWITHA BOOWITHA BOOWITHA

—It's over, Dad said from the doorway. Now give it to me. When you learn how to take care of things, maybe I'll let you keep these.

Two

—What do you wanna do, Corie?

—I don't know, whatta you want to do?

This is Karen's house. In the new tract, with dirt lawns, dust lawns, no sprinkler installed by fathers. With the hosewater making mud rutts through the yard. With no furniture in the den. With laundry all over. Where is her mother? With jelly on the kitchen counter. I'm hungry.

—I don't care, whatever you want to do.

—But it's your house, Karen.

Karen's house. Baseballs and mitts, little brothers, pieces of broken crayons on the floor, dump trucks, miniature gas stations, dogs on the couch chewing pillows. She is not confused.

—But you're the guest.

—But I don't know what there is to do here.

What *you* want to do, what your room looks like, how many sisters and brothers you have, what they do, where your mother is, why the piano is in the front hall. You wear Levis. Where is your father?

—Whatever you'd do at your house.

Play kickball, watch the guys across the street practice in their band, go swimming, look for rocks, search the shack in the field, sit in the field, paint in the field, paint, paint without looking, paint the reflections on the window, paint you, hide under the bed to paint, paint.

—But what do *you* usually do?

—I don't know, sit around, go to the ranch. Wanna go to the ranch and ride horses?

People in Levis. People I don't know. People who know how to ride horses. Hooves that step on your feet. Sweaty legs that kick, teeth closing together.

—I don't know how to ride horses.

—I could teach you.

—How would we get there?

—Hitch.

—I don't want to hitch.

Kissie. Chicken.

—So whatta you want to do?

—I don't know.

—We could go hiking.
—Where?
—The hills start at the end of our street, see?
—What's back there?

Snakes. Men with rusted scissors. Soggy wooden bridges. Barbed wire hidden in the field grass. Nails on end.

—Just hills.

Spiders. Old tires. Wild horses. Cow teeth. Burnt-down houses, remains of concrete foundations, remains of chimneys.

—We don't have to hike all the way to Twin Trees.

Rotting doors, spotted scratched stand-up bathtubs. Stopped-up creeks, mosquitoes, fallen eucalyptus, mud.

—I don't have the right clothes.

Shoes with slick bottoms to slide, thin pants to rip, short-sleeved shirt to scratch arms as branches flick back.

—I have something you could put on.

Tumbleweed. Rocks housing lizards and potato bugs. Fields of foxtails in my socks. Potholes. Slippery embankments. Dust, dust, dust.

*

There are four found trails to Twin Trees, more if we go looking, we will go looking, catch me go looking. We went down one of them, will go down the rest soon.

There are probably more. Soon. Tomorrow after school. Next Saturday after I clean the house. Right now after my feet rest. Right now.

The horses didn't bite, somebody owns them. They on the owned side of the NO TRESPASSING sign. I on that side too. On their side, with them, owned by them, by the inside of the fence. Disclaiming myself that tis the other, bending under the fence, out of the other.

The horses didn't bite, pet them on their noses like this, see? Scratch them between the nostrils, their tongues jump out. Tongues lick, not afraid, knowing I am, laughing nostrils at me, thinking I'm funny. Kissie, Chicken. Karen is not confused.

I am less hesitant now. Hesitant to bend under the fence, to disclaim myself that is the other, to trespass into my own, to bend out of the other. Soon. Catch me gone looking.

Karen is not confused by the places. There are places hiding here in these hills. Hiding places, something buried under, houses, someone, under these hills, in my direction, pointing. Following lines to the hills, to the dirt over buried places.

Into places hiding, there are hiding places. I will find them, find the secrets in these hidings. Catch me go. Olly olly oxen.

And there is a new silence—tell, they wouldn't—a filled space of silence running through these places hiding, that keeps their secrets, that keep mine.

I will bring home the doorknob, the knowledge of the hiding, of the buried. The brass doorknob, the burnt doorknob with the hole bending in, with the turning that doesn't tell of the secret.

Dad will help me polish it, silent, official, he will ask me why I bring home such a ridiculous thing, will humor me, will stare at it, crouched on my placemat at dinner.

He will not ask me to take it away. He will ask me when the fire was last, will wonder how old it is, will tell me it is too dangerous to go out there again, two young girls.

I will go out there again, with a new dance in my sneakers, with Karen who is not confused. I will disclaim myself, again, that is the other, will trespass into my own. I fill find them, the places, will find them hiding, will run silence through the places that keep secrets, that keep mine. Catch me go. Olly olly oxen.

Three

—Chuck it! Mouse is coming! Quick we're gonna be—
  DA DA DA DI
  DA DI DA DUN,
  DO DUN DA DUN,
  DO DUN DA DI DA DUN,
  DO DUN DO DUN DO DUN DA DI
  DI DI DI DA,
  DA DI DA DI DA DUN,
  DUN DO DUN,
  DUN DO DUN,
  DUN DO DUN DO DUN DA DI,
  DA DI DA DI DA DUN.

—SSSSSSSSHHHHHHH CLASS! That was the bugle now. Adams?
—Here.
—Afrant?
—Here.

—Shut up you guys! Did you hear—
—Barsoc?
—Here.
—Brandt?
—Here.

Mouse eats cheese.
—Carron?
—Here.
—Clower?
—Here.

—Gimme my pen, wouldjoo? Gimme my—
—Delaney?
—Here.
—Dunlap?
—Here.
—Elliot?
—Here.
—Errecart?
—Here.

(*How many inches above your knees, Corie? / Two. / How many? / Two. / How many? / Two. / How many? / TWO! / How many? / Two, MRS. PRINCE!*)

—Fenner?
—Here.
—Friedman?
—Here.
—Goodman?
—Here.

(*And how many inches is your skirt, Corie? / Four. / How many? / Four. / How many? / FOUR! FOUR! FOUR! / Four, Mrs. Prince . . .*)

—Harrington?
—Here.

I once knew a dog named Prince.
—Holt?
—Here.
—Ingersoll?
—Here.
—Johnson?
—Here.
—Judd?
—Here.

—Hey kissie, where'd you get those cute socks? YOU, kissie!
—Kane?
—Here.
—Kehoe?
—Here.
—Kraft?
—Here.
—Lear?
—Here.
—Can I have the hall pass?
—Why?
—I gotta take a piss.
—Lyon?
—Here.
—Macaulay?
—Here.
—Mendoza?
—Here.

—What are you so happy about? Your mom just died.
—I know it.

—Mr. Jacobs made us have a moment of silence for her before history class.
—He's a jerk.
—Chased Shelley around the desks yesterday when he made her stay after.
—I believe it.
—Goin' to Fabian's class today?
—Yep.
—Kissie.
—Fuck off.

—Moore?
—Here.
—Nolan?
—Here.
—Ocampo?
—Here.
Step on a crack you'll break your back. Step on a line—Ditch for it! The church parking lot! MOUSE!

—Osborne?
—Here.
Mouse eats cheese. I once knew a dog named Prince.
—Peck?
—Here.
—Pollock?
—Here.
—Rausch?
—Here.
—Rizzi?
—Here.
—Sanders?
—Here.

(*How many? / Four. / How many? / Four. / How many? / I ONCE KNEW A DOG NAMED PRINCE! / You're forcing me to call your mother. / That'll be interesting. / She works? / No. / Then you'll just sit here until we get a hold of her.*)

—Sherwood?
—Here.
—Squire?
—Here.
—Torres?
—Here.
—Tyner?
—Here.

—*You're Mr. Fernam?*
—*Yes, Corie.*
—*Who sent me here?*
—*Mrs. Prince. Is she still trying to get a hold of your mother?*
—*It won't work.*
—*Yes, I hear she's dead.*

—Underwood?
—Here.
—Van Benschoten?
—Here.

—*How did you know?*
—*Your father called us.*
—*You didn't tell Prince?*
—*She's busy calling. I wouldn't bother her, would you?*

—Ward?
—Here.
—Watson?
—Here.

—*No, I wouldn't.*
—*You want to talk?*
—*No.*

—Williams?
—Here.
—Wong?
—Here.

—*Then you can go back to class. I'll tell Mrs. Prince that I sent you.*
—O.K.

—York?
—Here.
—Young?
—Here.
—Zimmerman?
—Here.

—*You like to paint?*
—Yeah.
—*Jan Fabian told me about you. She said you're good at it.*

                ✻

    They are the same now. They have not stopped using mascara and getting too close to the mirror. The toilet stalls are still gray, they have not stopped being gray. The boys with pimples, with glasses, with stuck-out ears and math books, still huddle in the corner between the stucco wall and lockers.
    They have not stopped stealing pens, throwing erasers, borrowing notebook paper, they have not stopped making airplanes, chewing spit wads, shooting them through straws, off their fingers, off their knees, through the space between their front teeth.
    Looking, they are all looking at me now, and I am staring back as I would always stare, as long as they look, as she would have wanted.
    Shelley is still taping glossy men from magazines on the inside of her locker door, they are still smoking in the head near Mrs. Peary's room, the member of Girls' League still pretends not to notice, they are still not beating her up.
    Recess still changes to nutrition in the seventh grade, to brunch in the ninth, and they continue to have it, persist with that time around ten when we get restless and are left ten minutes free. Time when I will see Karen again, time when she will see that it is not the same.
    They are looking, all of them, staring back my malaise, my

growth toward some darkness that I have conspired with while I was away, while they were the same, they are the same.

Four

She is to you now, a bad smell, a queer noise, hair on the pillow, she is a bubble in our stomach, popping. She is in your plate.

Dad, why don't you ever talk about it? Pass the butter, thank you. *Tell us how it happened. Were you awake?* He is chewing. He is twirling the noodles tight onto his fork and he is chewing. Chewing. She is in his plate.

Did it take long? Cough. *Did she—* Pepper, get your paws off the linoleum. *Did she know it was gonna happen?* He makes sure Pepper gets off of her. Stay off. He returns to the table, twirling his noodles again. Squeezing her tight onto his fork. And chewing. He is chewing.

Did she warn you? He adds sugar to his coffee. *Did she say anything?* He watches her steam up and fog dimly against the window. Gail fingers a line across her. Don't do that. It is too hot to drink yet.

Was she in pain? Did she want you to tell us anything? He shuffles in his chair. We should get these chairs fixed. It tickles her. He reaches for the cheese, for her, so he can begin twirling again. Twirling her tight, squeezing her on the fork. He is chewing. She is in his plate.

Do you think we shouldn't know? Do you think we're too young? He does not swallow. He puts her down, still on the fork, rests her against the plate, tries the coffee again. She is still too hot.

How come you won't answer me, Dad? Did she leave us anything? Did she leave a NOTE? He burns his lips, she is still too hot. He gets up off her, she is laughing, tickled. He walks across her to the door. YOU killed her!! he screams, spitting her out, taking a breath of her. YOU killed her! YOU! and points her at me, breathes her in. You MURDERED her! With THAT! With those QUESTIONS! and he throws her at me, the full fork, the hot coffee, the wet window; he slams her hard, splintering the door jamb, he drives off in her, the engine grumbling: You, you, you murdered her you murdered, you, you, yoooouuuuuu.

*

My mother used to keep a gray smooth stone on the ledge of the bathroom window. holes in it. pumice. it will smooth the edges of your soles, here, try. drop the washcloth in the water. splat. soap in her face. put your knee down. splat. hands cold, pink, wrinkled. too long in the water. I always wanted hers. long, angled, green veins bulbing up like sculptures. she wanted mine. close your eyes. mine. smooth, fat, all the way across. always in my nose, yesterday's fort under nails. moons hidden. skin grown over them. bit them. pumice only for feet.

steamy here. white towels. don't touch me there. mountain peaks. put her in a trainer. hook it and then turn it around to the front. there. no, you can't sleep in it. tell her why she can't sleep in it. too constricting. feel safe.

white ridges now. never in my nose, never in the dirt. no fort now. green veins bulbing up like sculptures. cuticles eaten off. mine fat, smooth, all the way across.

gone now. both gone. mine. both mine. wanted them.

❊

The water is drip / is drip / is drip.
WHAT ARE YOU DOING IN THERE?
It is four in the morning. Gail is trying to sleep, is trying to listen, is trying is try / is try / is try.
ARE YOU READING?
Dad is standing on the other side of the door, the other side of the waking, the water, the other, is yelling, is not trying to listen, is not trying, is not / is not / is not.
ARE YOU SITTING ON THE COLD FLOOR?
Only the doors to the bathrooms lock, lock the water, lock the other out. The waking, the water, the other, is drip / is not / is drip / is not.
ARE YOU PAINTING IN THERE? GETTING COLORS ON THE SINK?
He is standing on the other side, standing in his bathrobe, he is calling me PECULIAR.
WHY WON'T YOU COME OUT OF THERE?
Peculiar to be alone, to be in the bathroom, to be on the other side, to be in the locked, in the morning, in the waking, in the water, in the water, the water.
WHY DO YOU CRY LATE AT NIGHT?

The water is drip / is sound / is shape / is splat / on the sink / is voice / is the book / reading.
WHAT ARE YOU DOING IN THERE? WHY ARE YOU IN THERE ALONE?

❋

The angel lights up. She is plugged in.
—You think you're getting what you want, Corie?
—I don't care.
To Gail, red and silver; to Corie, soft rectangle. To Dad, to Gail, to Corie. Boxes. Speak nothing, hold nothing, sense nothing, mean nothing.
—How come you're not watching TV?
—I never watch TV.
She is plugged in. She is at the top of the tree. She is white. She is smiling.
—Dad thinks you should be with the *family*.
—I want to sit under the tree.
Under the tinsel falling silver, the smell falling secret, under the angel falling sideways, the promise falling forgotten, the past falling deaf.
—Dad's letting me go to the New Year's party with Mike.
—I have to *watch* you, Corie.
—You want to string popcorn or hang stockings?
—Dad'll feel left out.
—He can too.
—He wants to watch the rest of the program. You wanna watch it with us?
—I never watch TV.
She is plugged in, is at the top of the tree, is white, is smiling.
—Dad thinks you're being morbid, Corie.
Is the sound, the shape, the splat on the sink, is the voice, is the book, reading.
—He thinks you should be with the family more, Corie.
She is in your plate.
—He thinks you're being morbid, Corie.
Is the speaking fallen secret, the sensing fallen silver, the holding fallen sideways, the meaning fallen deaf.

Five

My mother is making some sign to me, one long finger outstretched. She is pointing at me, at my hands, she wants to see my hands. I hold them up to her and she squints in curious amusement at my nails, filled with thick creases of yellow dirt, at my palms, lined and stained. She reaches for my wrists. Catch me.

You plan to eat at the table like that? If I were still there, you would not sit at the table! And she watches in horror my yellowed hands putting the napkin in my lap. She stares speechless at my hands folded for Dad's prayer. She paces through dinner, slightly irritated, waiting, but there is no remark.

I get up to leave and she is trying to grab my arm above the elbow, she wants to take me aside, and look closely at the new hairs growing at the edge of my temple. She wants to muse slowly over the shape of my face, and my hands. She pleads mutely for my hands. But they are no longer the fat smooth hands that she had wanted. They are becoming more and more like hers. Angled, the veins rising slowly to the surface like thin luminous strips of jade, laced awkwardly in brownstone.

I walk the hallway to my room and she grabs at me from the glass picture frames, from the open crevice of black in the bathroom, from the slit of air underneath Dad's closed door. She is hovering over the medicine cabinet as I brush my teeth, stiffly judging the thinned fingers that hold the toothbrush. Push back your olive skin! Make the moons show! She grins lightly at the idea, delighted at those tiny white moons she had given me, so carefully training my dark skin to reveal them. She did not give me my olive darkness. She was almost pale, she was like a flat stone, bleaching in warm water and thick sun.

The dog scratches against her thigh as she follows me into my bedroom. He fogs a little patch of trouser, he licks, then sniffs the familiar scent. His sense dulled by his own wet breathing, he leaves the idea, a tiny tongue-wet spot on the wall, leaves it suddenly, reminded of his water bowl and the smell of lightly cooked fish.

She sits with me at the foot of my bed, she looks with a quick disappointment at the walls that she no longer commands. But she is distracted by my yellow hands, clutching the drawing pencil, securing the pad on my knees. She wonders what it is like, this place with yellow dirt. She wonders if I were digging.

I will stare fiercely at the page until I no longer have her warm scent, till her body no longer moves the bed, till my muscles calm the urge to reach out for her wildly in the air.

She is very still. Her head bends just slightly so I will not notice that she is trying to inspect my knees. She sighs. I am almost too old.

Six

The door to the den is hollow. It must be, the thin wood splintered easily under the weight of his arm, and he is hardly rubbing his hand, or acknowledging Gail's scowls. Karen says they are all hollow. We shut mine without a sound, safe to watch them through the long vertical slit where the door does not fit properly into its frame.

How many months is she late? Two months. My sister is two months late. Sitting in the living room, she tucks a moccasined foot under her thigh. Keep your feet off the couch. So young. You are so young. We were going to get married anyway. He gets up to fix a drink. Oh Dad, why drink? Why drink! Why ruin the door. I am your daughter. We were going to get married anyway, she offers.

Will she remember? Under the white veil, under the white sheets, will she remember the hollow slats, the hole bigger than a fist? Will she remember the door?

A shadow is growing darker through the door. We plunge to bed and desk without a word, I securing a pencil, Karen an album cover. He opens the door without knocking, and he is there, staring, screaming: DID MICHAEL EVER GET INTO YOUR PANTS!? Gail is following, Dad! Dad! Do you really have to drag her into this? DID HE? DID HE? He is letting himself be dragged down the hall again, giving up his last attempt at control to that hand, fingers pinched tightly around his arm, just above the elbow. We watch them recede into the light.

Will she remember, clutching at the white dress, clenching wildly at her white hospital smock, will she remember the door? The hollow slats, the hole splintered bigger than a fist? Will she remember?

We close the door quietly against the gaping brilliance of the hallway, leaving only the long vertical slit where the door does not fit properly. Karen lays the album flat across her stomach. How many months? Two. Did you guess? We were not running out of tampons. She took long showers, she read my health book while I was at the movies. We were going to get married anyway, she keeps telling him. Are you sure this is what you want, he keeps saying. Do you know you won't be able to finish school? He is calm now. He will not do that again.

Later we hear the clanging of metal against metal. I peer through the slit. He is unscrewing the hinges.

❖

Dad is yelling at Gail's bedroom. Pepper is barking back. No one else answers. What time is it?

Dad is beating something against the walls in the hallway, pacing up the hall, the dog biting at his ankles.

What time is it? Is it morning? It is quiet now. Pepper whines and scratches secretively at my door. Let me in.

I let him in, and he waits with me patiently. He will not tell me what it is, the screaming, the silence. He smells damp, as if he has been out in the grass, as if he has been digging.

He barks. There are sirens. He hears them, sirens screaming, sirens, silence. He is at the window now, barking at the sirens, at the red and blue lights making leaf patterns on the closet door. He is barking at them, the sirens, these are the sirens.

What time is it? How old am I? I go out into the hall and Dad is glaring at me. He is wringing Pepper's sock in his hands.

—GET BACK IN THAT ROOM OR I'LL—

But I am too old. He lets himself be pushed away.

—IT WAS AN ACCIDENT! AN ACCIDENT! he threatens me, threatens the dog, the sock, threatens to leave by the window, threatens to be helpless. He has stood in the hallway before. He hears them.

—Gail, what happened?

—I'm sorry, Corie. I had to get rid of it. She points to her stomach, and hands me two thermometers, emptied of their mercury.

—You call that an accident!? I shove the two thermometers in his face. He grabs them and throws them against the wall. Pepper is growling at us.

—NO, GODDAMN IT! my father says and starts to cry, holding the sock between his teeth.

What time is it? How old am I? I am too old. I leave him there with his sock, Pepper follows me down the hall, they are knocking and I let them in, I point to the hall, to the bedroom, to my father; I hold the dog, he is growling at them, the sirens.

They are leaving with her now. Dad is leaning on the bedroom door, beating his head rhythmically against it. He clenches down harder on the sock with his teeth, they are taking her the sirens, these are the sirens.

—Can I come?
—How old are you?
—Fourteen.
—Yeah. Give the dog to your father.

He takes the dog, still growling, and sticks the wet sock into his mouth. Pepper does not want to play. They are taking her, the sirens.

*

He is sitting at the counter in the den, filling in numbers neatly with his colored pencils, labeling the placement of light fixtures, outlining display racks, placing cash registers. I am going to ask him now. He is going to remember.

—Dad, I want to talk to you about her.
—About Gail?
—No, about Mom.
—What about her? (Scribbling on his neatly placed figures, coloring the display racks, remembering.)
—I want to know what she was like.

He grinds the blue pencil in the sharpener, blows it clean, twirls the tip between his fingers, goes on working without looking up at me. He is not going to answer. He is going to play. Wanna play?

—I am old enough now that I want to know what Mom was like.
—We never fought.
—She sure yelled at Gail and me.
—She was worried about you.
—Gail?
—No, your mother.
—Why?
—Because you were the youngest, I guess.
—How did she die, Dad? (sirens screaming)
—A heart attack, I told you at the time. Don't you remember?
(sheee yadda yadda)

—Yeah. Did she leave us a note or anything?

—It happens too fast, Corie. Don't you understand what a heart attack is?

—Yes, Dad. (Don't hit me!)

—She said something to you when she was under.

—What?

—I don't know, something about you digging something up.

—Digging? (She wonders what it was like, this place with yellow dirt, she wonders if I were—)

—Something from the garden.

—We don't have a garden, Dad.

—We did then, it was right out there beyond the sliding glass doors. A flower garden.

He holds the tracing paper to the light to inspect the display cases. Wanna play?

*

I look at my mother through the glass of the door and I wonder why she is sitting there, in the wet grass, so serious, the dog attentive next to her, both of them, staring at the yellow flowers in the garden. There are still stars out.

Her face is harsh as she first sees me standing there in my robe, spying on her as if I have caught her in some secret act. Then her face softens, her long arms push her up off the moist ground, she wipes her hands, her knees, and comes to the glass door.

—What are you doing, Mom?

—Go put your coat on and come out here, Corie, I want you to watch something.

I run obediently to the coat closet, opening it carefully, listening for his snoring. This is not his secret.

I kneel down with her, facing the flower garden she had planted a week before, the tiny yellow flowers drooped in sleep. I am about to ask her but she senses my bewilderment, and touches my arm lightly with her fingers.

—They are going to wake up, she tells me. They are going to wake up. She looks at me, content now, her eyes tired. She will sleep soundly after this.

So I wait with her, in the garden, the dog breathing fog patches next to us. She looks up to see that the stars are fading, and then she watches the yellow flowers open one by one, holding me tightly as if I might vanish, until they are all standing open.

—There now, you see. They are all awake, she tells me. She looks down at me, and I see that she is crying. Let's go back to bed now. A twelve-year-old girl shouldn't be awake at five in the morning.

She leads me back to bed, setting my teddy bear against my stomach. She sits down for a moment, wiping her eyes.

—Don't you wonder what's under there, she asks, to make them wake up like that?

And I tell her it is the sun that wakes them up, not something buried.

She smiles at me, as if to tell me how young I am, how small I look in that bed, to tell me that she knows better. She kisses me.

—I'll let you sleep in this morning, she says.

❊

Her memory has wakened me. I inspect the thin lines that trace along the ceiling, imagining myself to be traveling them like roads in the desert. I force myself to remember all that will bring my head back to the pillow; I tear my eyes from those roads and make them lines on the ceiling. I am fourteen, she is dead, and I begin to come back to the early mornings. I leave that other morning that has wakened me; I roll onto my side with a groan, as if to forsake that memory. But the pillow cracks under me.

I do not believe in coincidences. This is not *his* pillow. I am fourteen. There is a note there. I look around the room at this morning, at my old teddy bear, abandoned, in the corner by the closet; I listen for the familar noises. There are none. I am indeed fourteen. But the note is still there in my hand, and the memory of the cold dampness of the morning still clings to my back. I do not believe in coincidences.

Dear Corie:
Meet me Sunday morning at five in your yard by the glass door. Bring something of yours that is special. I will bring something too. We will bury them. Then whenever we see the flower we will remember what is buried, and we will have the knowledge of the secret.

Karen

—Did she leave us anything? Did she leave us a note?
—I don't know, something about you digging something up.
—Digging?
—Something from the garden.
—We don't have a garden, Dad.
—We did then, it was right out there beyond the sliding glass doors. A flower garden.

I am jumping out of my fourteen-year-old bed, the garden! The garden! screaming toward the garage to get the shovel. The garden! It's in the garden! and I am out there, in the morning, Pepper sitting patiently next to me, digging into the wet grass by the sliding glass doors, digging, into the grass where the garden used to be, next to the door, where I stood watching her, as if I had caught her in some secret act. The garden! I look to the window, expecting to see myself there, twelve and curious behind the glass. But *he* is there, staring.

—CORIE, WHAT IN GOD'S NAME ARE YOU DOING?
—I'm digging.
—IT'S FIVE IN THE MORNING! YOU'RE TEARING UP THE GRASS!
—This is where the garden used to be.
—GET IN HERE! YOU DON'T HAVE ANY SHOES ON!
—She left me something in the garden.
—YOU CAN'T DESTROY THE GRASS LIKE THAT! YOU'VE GONE CRAZY!
—You know Mom left me something in this garden.
—GET IN THIS HOUSE!
—Don't touch me. I'm digging.
—GET IN THIS HOUSE BEFORE I DRAG YOU IN! THERE'S NOTHING THERE!
—Did you take it? Let me go.
—DON'T YOU HOLD THE SHOVEL AT ME LIKE THAT! THERE'S NOTHING THERE!
—You took it then. She left it for me. Let me go.
—THERE'S NOTHING THERE, CORIE! STOP HITTING ME WITH THAT SHOVEL! YOU'VE GONE CRAZY!

*

There's a white plastic box next to my arm, with buttons on it.

I press the first button. The back of my bed raises methodically. I notice that the sheets are white, the blanket a pale peach color, and that the shiny metal railing gleams my face back to me as the bed moves down again.

Next to my left arm is a little rolling table. A plastic cup decorated with bending straw and a turquoise-blue plastic pitcher are sitting on it. On the other side of me, against the far wall, is a sink, and another plastic cup. The curtains are closed against a long thin window.

—Corie Barsoc?
—That's me. Where am I?
—St. Mary's. In the Mental Health wing.
—Mental Health wing?
—Your father brought you in.
—Because I hit him with a shovel?
—Nervous breakdown.
—I see.
—Is there any reason why I should feel like lead?
—That's probably your medication.
—Medication? Right.

I look around the room a second time, at the plastic buttons, the peach blanket, the metal railing that trims the bed, the plastic cup, the bending straw. St. Mary's. Medication. Don't be afraid. I hit my father with a shovel. There's nothing there. In the garden. St. Mary's. St. Mary's!

—Corie Basoc? Dr. Rucker. How do you feel this morning?
—Like lead.
—That's probably your medication.
—What was it for?
—To calm you down.
—To calm me down.

Plastic buttons, peach blanket, metal railing, plastic cup, bending straw. St. Mary's. Medication. Don't be afraid.

—Huh?
—I said that there's no need to be afraid. We're here to help you.
—Right.
—Could you tell me why you hit your father over the head with a shovel?
—Over the head?

—And what you were digging for in the lawn at five in the morning?
—Dr. Rucker, is Dr. Forrester still here?
—Yes, in emergency, why?
—If you'd bring him in here I think I could explain more quickly.
They are mumbling in the hallway. Voices in question. I review: plastic buttons, up . . . down . . . face in the metal railing. Soap in her face, don't touch me there. White sheets. Will she remember? clutching the peach blanket, plastic cup, bending straw. Don't be afraid.
—Dr. Forrester?
—Yes, Corie.
—Do you remember working on an emergency case about two years ago, a suicide of a Rebecca Barsoc?
—Yes, I remember, Corie.
—I said, a SUICIDE case.
—That's right, Corie, we were just talking about that. It was your mother.
—Right. She killed herself?
—That's right, Corie. She killed herself.
—Why did she kill herself?
—Well, I think this is a family matter, that you should ask your father, what do you think, Dr. Rucker?
—I think she needs to be told, Dr. Forrester.
—She killed herself because she was dying of cancer, Corie.
—Dying of cancer.
—Why don't you lean back against the pillow, Corie. There now. That's better. Now Corie, tell me about the shovel.

Don't you wonder what's under there? to make them wake up like that? It's the sun that wakes them up; not something buried. Digging? We don't have a garden, Dad. We did then. Wanna play?

Plastic buttons. Let go. let go, I'm digging. You took it then. metal railing. Over the head? White sheets. Peach blankets. Mom left it there for me. Sheee yadda yadda. There, there now. Nothing there. You took it then. YOU! Plastic cup, bending straw, medication. Don't be afraid.

❊

—I fixed you breakfast, Corie.
He is standing there, the spatula in his hand.

—You want breakfast?
—I've had breakfast. Why are you wearing that thing around your waist, Dad?
—To keep me slim.
—His ribs are broken, Gail answers.
—How did you break your ribs?
—When you hit me with the shovel.
—That's cruel, Dad! My God! Tell her the truth!
—I fell in the shower.

Dad stirs his coffee, his eyes on the window. He is chewing slowly, and stirring.

—Guess who I saw at the hospital, Dad?

He butters his toast.

—Dr. Forrester. I asked specially to see him.

He offers me half.

—No thanks. Do you remember him, Dad? He saw Mom when she committed suicide.

He bites into this toast. He chews.

—I said he saw Mom when she COMMITTED SUICIDE. You were there. Don't you remember?

He offers me more scrambled eggs.

—Dr. Forrester told me that she killed herself.

He takes some scrambled eggs.

—Because she was dying of cancer.

And some bacon.

—DYING OF CANCER.

He offers me bacon.

—Dr. Forrester said he tried to tell you, Dad, but you said you knew already.

He looks out the window. He stirs his coffee.

—Apparently they'd been treating her for cancer.

He offers me coffee.

—You know I don't drink coffee, Dad. Anyway, he said he also heard her tell me to dig up the—

He pours the hot coffee into my milk glass. The glass breaks. He continues to pour coffee until Gail takes the pot from him, and runs to get a towel. She wipes the table.

—That she was asking me to dig up the garden.

He butters another piece of toast. Gail watches closely from the hallway.

—What was in the garden, Dad?

He offers me a piece of toast.

—No Dad. What was in the garden?

He slices the butter for me, and lays it on top of the bread.

—What did she leave me in the garden?

He looks at me, and begins to butter the toast.

—WHAT DID SHE LEAVE ME IN THE GARDEN, DAD!?

His arms go flying, the buttered toast against the window, the knife sticking into the wood of the kitchen cabinets. His chair is knocking against the wall, and I watch him run across the lawn screaming Rebecca! Rebecca! and flailing his arms. Rebecca! He beats his fist against the car window, Rebecca! finds the keys in his pocket, Rebecca! and he is gone.

*

Pepper is not sure why the heavy-smelling Indian rug has been wedged between the seats of the car, why the clay sculptures under the bureau, spotted occasionally by his wet interrogation, have been wrapped in newspaper, and stuffed into a box. He follows me out to the car, propping his front feet up on the fender to inspect the suitcase, the stereo, the crate of plants. And why did I get to sit on the linoleum during dinner, he seems to wonder. I shut the trunk. He will miss those dust smells, the dirt crumbling onto his tongue when he licks the clay. But he will go digging. Sometime we will go digging again, o.k.? He does not believe me.

He follows me into the house, we watch my father pull up the white projection screen and hook it into the stand. He unscrews a lens in the old blue-gray movie projector.

—What are you doing?

—There's an old home movie I want you to see before you leave for boarding school.

Pepper gets nervous. He patrols around the coffee table, knocking books and vases as he goes. Distract me. Read to me. He sniffs my dictionary of verbs. Read to me.

—To go. Going. Gone.

Yes, this is what it is. He sits down.

—I go, you go, he she it goes; we go, you go, they go. Or: I do

go, you do go, he she it does go; we do go, you do go, they do go.
He barks once. Continue.

—I am going, you are going, he she it is going; we are going, you are going, they are going.

My father starts the projector. Uh-oh, keep reading. Gail sits down next to me on the couch, tucking her foot under her, nodding her head lightly to the rhythm of the words.

—I was going, YOU were going, HE SHE IT was going; WE were going, YOU were going, THEY were going.

The screen turns to mauves and grays. We see fingertips extended, then two hands, then long arms sweatered in beige. I run to my mother, and she scoops me up as if I were a cat.

Pepper barks at her.

—I went, you went, he she it went; we went, you went, they went.

She puts me down feet first and I waddle over to my sister. She dissuades me with her hand. Don't touch me.

Pepper licks my arm. Keep reading.

—I will go, you will go, he she it will go; we will go, you will go, they will go.

I start leaning backward, and the long hands slip under my arms to catch me. She stands me up sturdily on my feet. There now. Go on.

—That I may go, that you may go, that he she it may go; that we may go, that you may go, that they may go.

I walk over to a hedge of ivy, the hands following behind me, and start to walk along it, following its clipped surface with the palm of my hand.

—That I might go, that you might go, that he she it might go; that we might go, that you might go, that they might go.

I move away from the hedge and a little hand taps me on the shoulder. Wanna play? catch me.

—You're it! Gail screams from the couch. Pepper barks.

Before I can turn she has hidden behind the pear tree at the back of the yard. I turn to my mother. No, she nods, not my hand. She catches my elbow because I am tilting. She whispers in my ear. Go looking.

—I would have gone, you would have gone, he she it would have gone; we would have gone, you would have gone, they would have gone.

My father is snoring on the couch. Pepper barks and heads for the sliding glass door. It's only a movie, Pepper, sit down. Gail returns to my book.

I look toward where she has pointed, between the two pear trees. There is no one. I walk in that direction, then stop. I turn to look. No one is following me.

—That I might have gone, that you might have gone, that he she it might have gone; that we might have gone, that you might have gone, that they might have gone.

Gail wakes up Dad, and asks him why we are watching me waddle. This is how she learned to walk, he says. He goes back to sleep.

I am heading for the trees now, tilting slightly. Gail bursts out from the other side, screaming and flailing her arms. She runs down the hedge to the patio, wrapping herself around our mother. Olly olly oxen, Free! Free! Free! Mom pats her on the head and straightens her sweater. They look to me, and my mother holds out her hand, fingers outstretched. The fingers recede, the hedge blurs, and I am there, walking, tilting, my palm flat against the clipped surface.

Go, let us go, go.

EYE BLADE

GEORGE EVANS

In the years after the war earth kept shaking. Landscapes shifted, windows shattered. It was unpopular to be unpopular. Strong words disappeared like bees into hives. Films were made, songs written, clubs formed, monuments erected. Objects slipped in and out of the sacred, in and out of the picture. Everyone paid to wait to get in then paid again and maybe got in. Cliffs dropped, mills shut, pots fell from windows, cars from bridges, hearts fell like coins through grating—earth kept shaking, wind rising. Weapons replaced gods, instruments people, TV replaced distance and the mind. The missile lines grew long, long, long as a child's breath long.

Air hazed by seed and bug, arms heavy
after work he sits, denting a can with his thumb
on the steps of the house which owns him
watching a figure inch uphill across the valley.
Skateboards scrape, spark, evening news begins
its tally, its explosions, the world exploding
block by block towards him, thing so large
the whole will not be seen.

Up blacktop as up a wall the figure climbs
cobblestone knuckled street, toting a sack.
San Francisco glitters in a circle below.

Hunchbacked splinter on a hill, inebriate
wandering, bottles clanking, moving but still:
dot, insect, man, instrument, subject.

A shadow. It peels and approaches
gathering shape: edges emerge,
throwing-star whizzing, it grows,
wheels away dragging its shadow—
a hawk, holding the world
in its eye: an island.

A blind man, stick tapping, stops, jerks,
rapping, tap
        tap, fishing, reaching, scraping
ahead, edge against edge, feeling the world
through his stick, the whole black sphere
attached to its tip.

A child in the shape of a bowl.
A powdered form, body striped with ribs,
eating dust without protest
in forced geophagy, eyes shut,
no world beyond its plain
littered with blanket shacks.
Stone. Ripened fig. Speck in the cosmos.
Pair of eyes like planets.

A black mark bouncing through forest
leaps over root and vine, ferns snapping
between her toes she runs from a village,
whose life is food, who longs for the north
with its malls, who covets abstract visions
of Texas, hears gunfire and looks back thinking
of a water pot on her table flowering as it bursts,
and of TV, if its figures, like water, are stored within.

A farmer marching to a cadence beyond the trees,
chanting, fluting, shirt frayed, dissolving through hills
to revolt, to reach the capital grown in him like a fist,
ignorant of what he'll face: that everyone wants control,
and no one wants control—what everyone really wants is money.
His voice is smaller than a pin's tip, a bird's beak
drilling air, waving its words like a flag,
a target lighting its center.

The figure moves deeper away through the hills
wandering farther, mapless, sighting blue and gray
peaks to measure how far, or remember nothing
is far, everything far: arriving isn't the point,
the point's to move not away from or to, but constantly
in the place where the mind centers,
driving it through mountains.

The road becomes a dot. Red-winged blackbirds
stick to the wires then explode across the field.
He's walking in extreme states through the landscape.
The beauty is painful. He recognizes long-denied voids
spanned or camouflaged for love or attention. He's stuck,
confused between grief and self-pity, knowing patience
can turn to bitterness and vanity greed, but what
of aimlessness and sentimental ruth which have
become his sack, and what of the journey
which has become his destination?

The poet Tu Fu, grain of dust to China,
watching chaos cross the horizon, pain
real and imagined, pities himself, hair
too thin to tie back, wanting his art
to bring fame, lost in the dead Li Po
drowned grabbing the moon where it blazed
like a white leaf past his boat, Li Po,
who drunk could make a ragpicker king.

Its nature is that it's outside, outside the outside.
Not a vehicle but a motion. No meaning, no correlative,
no use, no rejection, no acceptance, no form, no intention,
no morality, no religion, no school, no forebears, no
value, no price, no time, no bones. It is its own
future and purpose, own audience and shill.
It has no nature. It has the world.

Crowds: unanchored islands.
 Trees: loci underpinning land.
Stones: silence continues.
Ground: departing map.
   Sky: motion's room.
Animals: revolt subdued, and
 Wind: the mountain finally arrives.

Plowing uphill the runner blurs
the world, losing sense of the motionless, everything
moves: islands of flattened gum, discarded fruit,
exploding paper, statues jumping on lawns,
fences and clouds stream by. No distinguishing
what's forced to move from what itself is moving,
every speck of the world falling apart,
gathering, falling apart.

A mockingbird in its tree, long tail flicking:
guitars, cork pop, telephone, jackhammer, nail
squealed from wood, snoring. Creature of sound:
door squeak, hose spray, cop whistle, siren, jazz:
picks what has passed and sings back
over power line, building, car: voice
of the times on a twig, rocking,
returning what won't be had.

The Wall, the black wall rising. Dead list
in the capital, black list, stone mirror

faces float across searching its columns,
names to touch, lean upon and fall through
into space. Rolling stone which will not roll.
Wound which doesn't close except in sleep.
Numbers small for war, somehow unforgivable
for something perceived as error, though all
battle is all error. And when its shine dulls,
its sting fades, and those who weep go dry,
what good will be a wall?

An orchid in light
its veins are bones
the world is mad we
live on air unwinding
and resist the dark
but touch the dark
for death's the stem
and root, impatient
growling thing.

Bosatsu of a thousand whirling arms,
windmill on a hillside, each arm an event,
skill, perspective: world of a thousand
eyes viewing a thousand things at once
and each thing equal; world of a thousand
windows opened on as many things meaningless
one by one as there are self-obsessions;
the world which will not tolerate one view.

The obsessed turning in circles, gathering
chips as if this is Reno and somebody wins:
*I'll take one of those, one of those, and
one of those. The rest of you take a hike,
get fucked, bug off, hit the road corpse—
this is my movie, this is me and me me me.*

The Master said, getting up from where
he'd been kicked across the casino through
sawdust into a corner under a slot machine:
It's a bridge, all road and no sidewalk.
Don't fall asleep.

A flying ant crawls over an ash, stops,
washes its wings then goes on, lives to live,
exceeding neither limits nor potential,
excoriating nothing in its pursuits.

A boy in a tree at night above Main Street,
Peru. Below, a woman with a sack of sugar
on her head waddles to a traffic light and waits.
He has entered the world of the dead, and hears
drums on the ground below, drums lined up in rows,
each with a drummer: enemies of the Inca,
gutted then stretched in a smokehouse
until their stomachs are taut, fit
to drum through the jungle.

Inches above water the osprey hangs,
then dropping locks its talons in,
but the fish is strong, heavy,
and drags it into the liquid sky
through rings of impact down
to silence.

An old woman wobbles up, air bending
as she bends, wind pushing her down,
each moment stops, each step
an arrival she takes the hill
hobbling, teetering and suffers
the incurable: her skin.

A clump of weeds breaks the pavement
throwing lower worlds to light.
The rough the unwanted struggle most
but last by the effort, translating
land into landscape, revolting
underfoot. Things which are deep
and will not be cut.

Sparks fly from a guitar
player's fingers in a Chuck Berry duckwalk
across the skyline, wind shears
him, wind harp blown to life
on a high ridge culture dances,
twists, the passions course,
blood pounds and words drop
like hammer blows. Rock & Roll.

It's the specter of death. More than a cloud, a second
sky; Rangda, breasts like hanging knee-socks, shaking
her ass, whipping a trance on reason, stabbing the world
in full view of the world; Kali, popping her cheeks
like a drum, dangling life by its skin, flaying,
peeling the planet, palms out, hips in frenzy,
watching the world blow apart
as if on a screen.

In slo-mo the column rises into a ball, sweeping
cows from their fields, babies from cribs, hands
from their bodies, charring wallpaper, melting
the fence: no more flowers, trees or lawns, no more
houses, cars, or baseball. No more music, no more
sex, books, computers, airplanes, insects and birds
a vapor, elephants fried to a crisp, whales thrown
aloft: the reverse of the world.

A voice struggles to be heard,
but which mechanism detects it as a lie,
and how is it recognized? Patience
is the power of the eye, naiveté
skill when it comes to observing men
not given to truth, as silence is skill
in the wilderness, watching. A spot,
a mirror in light among mirrors, the eye.

Off your high horse, mister.
Yours is no profession or branch,
and you no bird to piss down
on fields and peasants, there are
peasants only to the stupid
who know prices, sizes, ways
to squeeze through, get in,
suck up, and nothing else.
Yours is not to perfect,
but to know you can't,
and not why.

The evening shakes
and glows on the bay.
Buildings, colors, people
gild the moving sea
pours away then back
holding all in one shimmer, one
wave, as the world itself will not,
which slips away, then slips away.

FOUR POEMS

FRED MURATORI

THOUGHT LEAPS ON US

in ambush. Frames and doorways where we wanted
only sea, a plotless desert, ambiguity kept distant
as the signal science measures from a nebula as yet unproven.
Now the innocence we claimed in spite of monolithic proof
it never was has truly fallen, lengthwise, on our heads.
Yes, this is and yes, we are and yes, we must do something.
That constant clanging first thing we awaken to: *must do,
must do, must do.* Our children see it in our eyes and ask
why they were born, what they were given life to do and
when, goddamnit, when, as if their small, accreting brains
were working toward a deadline. Was there ever uncreased
birth? Ever presence without rent? We pay our lifetimes
through for damaged merchandise, furniture and explanations
tried and used and repossessed. Lies make sense but truth
flies upward through our hands no more than wind or heartbreak.

(The title comes from a line by George Oppen.)

HUBRIS

The sorrow hand in hand with blind belief, the ending
placed where everything begins, the primal fear

of touch becoming love becoming guilt becoming
hatred, the hard collapse of ceilings, shrug of beams . . .
One day we decided we would stand, and dream, and take
our senses with us, withholding awed reaction to
the earthly things that live to be perceived as certain,
irreducible slabs of matter upright in our paths,
daring us to disacknowledge them and with them our own
selves. We did, and spun new spheres of meaning centered
on the noises we had molded with our tongues. It worked
until the meanings split and multiplied, carried us
like rivers gone amok with rain. We likened them
to rivers, and in the instant of that reference fell
like birds, which wakened to their names and fell.

BACHELORS

Devoid of daughters here, for each is fated
to the tabletop as though to white Moroccan sand
and heads lift not to light but only shadow,
passing motion just a glance against the iris
which washes out to rim unfeelingly, mechanical,
the ball-and-socket block-and-tackle clockwork
meshing at a level so interior and unsurmised few
instruments can plumb its measure. Listen flush
with them to no essential music but the brush
of silver knives on china, sighs well-spersed
and nectar slow. Admire linen suits unchinked
by accident or *faux pas* at the velvet rope. Say
nothing on your way home of the love remaindered,
of air that seems euphoria spread thin enough
to pass for struggle, or struggle that has ended.

THE APPEARANCES

Time takes shape or seems to only as it disappears:
a tulip lost in fields of matted weeds, something in
the eye that exists only when the weeping starts,
when other, better selves work free of sublimation,
telling us they could have, *could have been* had we become
them as we'd wished, before the wishes stopped and turned to
day on day on day of movement for the sake of movement,
an army of alternatives dismissed for want of interest.
The layering continues, the coarse protective strata
build until the earth is no more than a muffled
pied-à-terre equipped with beds and mirrors, shades and
dead bolt locks, its clocks all hurrying to start again.
Some people have their memories embossed or bronzed or
published as if true, still fish poured from buckets over
ice for our consumption, so alien we think they move.

THE BLACK FLAME

NOBUO KOJIMA

Translated from the Japanese by Lawrence Rodgers

1

I've decided to write a letter to you, Hiroshi. In the end, perhaps, I won't send it. Be that as it may, however, I have to write it. If I'm to write anyone a letter, you're the first person I should write to. Because for these last two weeks it's as though I was put on earth for you, experiencing for the first time in my life the bizarre, and this has forced me to think as I never have before, the object of my thoughts being you, just you. You'll understand. You, especially, will comprehend my feelings. Or rather, I think you ought to understand.

You work in the same office with me. Again today, I mutter to myself, my thoughts pursuing you as you cross the room.

You sat down at your desk just now. You're lighting up a cigarette. What will you do next? You put down your cigarette and—ah—stand up and loosen your belt a notch. Pretty soon you'll comb your hair back. Now you're looking out the window. It's raining. Next you walk over to the umbrella stand in the corner and lift your umbrella off the base, where some water has pooled. You let the water run off your umbrella and look for some place to put it down. In the end you hang it from the side of your desk. You then empty your rather full ashtray and look at the man next to you. He's someone you play *gō* with. You start talking with him, and for a while you're lost to any- and everything. Suddenly you turn in my direction. We smile and exchange nods.

When you turned and looked at me and smiled, I wondered for an instant what I should do. I was able to smile calmly. As though nothing were amiss. After I had finished showing you my smiling face, I realized immediately you were going to resume your earlier position, and that nothing could be wrong, that if something were wrong, you certainly couldn't expect me to be able to smile so naturally at you. That's it exactly. I even feel a closeness to you. There's so much I want to say to you. You finish your gō chat at last and begin working. This is what I'm thinking: *I'll bet you're going to tire yourself out if you don't sit a little further back in your chair.*

You go to the restroom. You have to go there often. You set off, shuffling. I give serious thought to following you, but leave for the restroom after you've returned to your desk. I walk where you've walked. For some reason I find this pleasant. I relieve myself where you've gone. You'd been there two minutes earlier. I'm standing now just as you had stood.

Your discarded cigarette butt bobs in the drain like a drowned man, caught in the force of new liquid. You smoke Hikaris. This is the Hikari that you smoked. I've also switched to Hikaris recently. I learned at your place that you smoke Hikaris.

You appear to be unaware of that. You're in a good mood today. Which is why you invited me to lunch. I was in a quandary: should I order the same thing you were having? Before I could decide, you ordered for both of us.

"This should do it, don't you think?"

"Of course. I believe that will be fine."

"You're an interesting fellow," you laughed.

"How is that?"

"Nothing. It's just the way you answered," you said, then: "My wife suffers from indigestion. She's been like that since she was a kid. When her stomach's bothering her, she's no good at cooking and restaurant food like this doesn't appeal to her."

Your wife Rumi's the kid sister of a college friend of mine. I realized this after I began working here and heard about her from you. How can you possibly claim knowledge of the condition of Rumi's stomach when she was a child? She was twelve or thirteen then, wasn't she.

"They say not eating is the best thing for indigestion."

"The trouble is she doesn't eat. It'll affect her health if she doesn't. She's headstrong and won't go see a doctor either."

I didn't answer, but watched your hands as you ate your lunch. They're big hands. Not surprisingly, you really have a splendid way of eating. And I know that you have big feet. As I stare at your mouth and hands in motion I suddenly wanted to tell you something.

"What is it?" you asked, turning your smiling face to me as I struggled to regain my composure.

"You ought to learn how to play gō too," you said.

"You can pass some of your work on to me, if you like. I have time to do it." As usual, my language was more deferential than yours.

"No, if I'm going to have someone do my work, I'll take it home and put my wife to work on it. I'm a firm believer in doing what one is supposed to do."

Well said, indeed. But you're not doing what you should be doing. Or do you know that?

I put in a good word for you—in a roundabout way—when I went to the section chief's office in the afternoon. He was an upperclassman of mine at college, remember.

"He's a good man. I'm not sure how exactly, but I'd just say he's good. I would not credit anyone who says otherwise."

I suddenly realized what it was I was about to tell you when we were having lunch. I wanted to tell you then the one thing I shouldn't tell you. Whether you're aware of it or not, I wanted to tell you myself, and yet it would seem you're the one person I shouldn't inform. No, you probably don't know.

And yet as I come to feel closer to you with each passing day, it's as though you've begun to share my identity. It's not merely that you're important to me.

This is how it is, my friend. Are you ready? We're going to have problems unless you're aware of what I'm doing and what my motives are, because, my friend, this is unlike anything else.

I can even see you smiling at me and saying, "What do you mean 'unlike anything else'?" That's when I would have the unbearable urge to come out with everything, once and for all.

Anyway, it's a good thing I didn't say anything then. I can't let myself eat with you. I just can't let myself eat the same things you eat.

What would I blurt out? I can see it now. It would be, *for no good reason, merely giddily hollow, and there would be nothing more.*

That's it. And that's how it would be. I'm a stranger to pain and have even less traffic with pleasure. The only problem is you. What I want to tell you is that you must not eat a meal with me or have a friendly chat with me, and commute to the same workplace and talk of gō, that these things cannot be.

Wait, maybe these things are all right. Maybe that's what you should do. It's just that I wanted to tell you, *to have your advice about,* the fact that *I have you on my mind.*

No. It's more straightforward than that. I want to tell you how I'm giddily hollow and that, at the same time, only this state of mind is akin to anything like pleasure, even though *that business* gives me no pleasure whatsoever. The true nature of this pleasure, that's what I'd like to talk with you about, and, at the risk of repeating myself, the fact that you're the most appropriate person to advise me, and the most inappropriate.

Each and everything mentioned above I wanted to address to you.

I doubt you know anything about it. I'm deliberately writing as though you don't. That's because you're involved. It's a matter I can't relate directly to you in a way you can immediately understand.

You remember when Rumi brought your daughter to the office early this evening? When little Yukie went off to get you? Just at that moment I had come out of the section chief's office and ran into Rumi at the elevator on my way to the bookstore. They say you're meticulous about her appearance. She's petite, and her thin, delicate chiseled dark face, nicely set off by her navy-blue raincoat, was most attractive, I thought. The instant she caught sight of me I knew I had nothing to say to her. I knew that if I did say something it would be a lie, that tonight we would merely do again what we'd done before.

She sent Yukie on to the reception desk as I approached her.

This was her chance to draw next to me, take my hand, and squeeze it. I returned her squeeze. She stifled a soft scream.

"You're coming tonight," she said, speaking rapidly. "To throw him out. I came to remind you."

"We can't stand here like this."

"It's such a long time to wait."

"You mustn't break your promise. You shouldn't come here."

"I know that. But you're where he is. All right, step away. He'll be coming."

I jumped into the elevator and started down. What a ludicrous figure I must have cut, I immediately thought. Am I not like someone playing the villain of the piece? I looked up and saw Rumi's legs in front of me. A door opened, and there you were, smiling benignly. I was filled with good feeling when I saw your smiling face. And yet how could you possibly smile? Now your smiling face angered me. Are you powerless to know what reality is?

Before me was an expanse of wall and windows. When the elevator stopped and I stepped off I suddenly felt light-headed and leaned up against a stone column.

If I in fact hand over this letter to you, you'll then know all there is to know.

I'm having an affair with Rumi. Well, I'm not really having an affair. I suppose from your viewpoint, however, it can only be called an affair. And it's precisely for that reason I have to talk with you.

Remember when I said a little while ago I had squeezed Rumi's hand hard after she squeezed mine? I felt absolutely not the slightest desire for her then. It was you I wanted.

I leaned against the column to get the better of my giddiness. I knew there was certainly no one who'd concern himself with this damned feeling, that I'd have to deal with it alone, and yet it seemed to me to be really unfair that only I'm made giddy.

You might angrily accuse me of irresponsibility, but bear with me. If you don't know what's happening, it's only natural you'd put me down as irresponsible.

You may think, perhaps, that I was lying in wait for Rumi, having deceived myself into claiming I was calming my lightheadedness. However, that wasn't why I was there. Far to the contrary, even there I was thinking of you.

Understand this. Nobody else can control us. You're the only one who can do it.

I went into the bookstore and looked for a book on adultery. First of all, I hadn't the slightest idea which shelf such a book would be found on. Should I look under novels or amongst the law books? In any case, I wanted a book describing with complete candor how a man feels about it. Ultimately, however, my search was fruitless. What I was looking for was to be found nowhere, and as for Western works, I had a pretty good idea from the movies I'd seen. One's passions are aroused, and because one's passions are aroused, one must submit, mustn't he, to God's law, for one lives in awe of God. In my case, I'd say you're my god. God with a smiling face. Those fingers, those feet, that body. You, the hearty eater, with whom I've supped.

When I looked up I saw that Rumi was starting home. Little Yukie caught sight of me and pointed me out to her mother. I looked at her face through the rain-streaked windowpane. I turned away. Moments later she approached me.

"Why are you looking so glum?"

Yukie was leafing through a child's picture book.

I looked at Rumi. *She has no reason to reproach me. There's no reason why I should stand for her faultfinding.*

She peered directly into my soul with those big, beautiful eyes. Peer though she might, however, she hadn't a clue. With that expression on my face, it went without saying she didn't even understand that she couldn't understand.

"It's nothing. He says your stomach is bothering you."

"It's your fault. Listen, do you understand? You're going to fall in love with me."

She deftly flipped over the book I was holding.

As I stepped outside with her we once more squeezed each other's hands within the folds of her raincoat.

2

I'd had a phone call from Rumi two weeks ago. She said there was something important she wanted to talk to me about and that she wanted me to come over that night. I asked her what she wanted to talk about, and she responded she couldn't tell me on the phone, but that I shouldn't mention her call to you.

I looked at you as soon as I returned to my desk. What does she want to talk about, telling me to come to your house and not to let you know about it? You go out a lot to play *gō*. Could this be the reason for Rumi's dissatisfaction? I'd met Rumi once at your place, but it was you who had taken me there. That was a good year ago.

I'd no reason to decline her invitation, so I decided to go to your house. It'd been a company house, but, curiously enough, you bought it. It's at the top of a hill. You go up the sloping road, and your house is by itself at the end. Rumi heard my footfall and looked out the window. It's an odd house. I knew as I walked along that anybody who was in the house would know I was coming long before I got near it.

But the fact is, you weren't home. You'd come down the hill a good deal earlier. I was now walking along this same street in the opposite direction.

You weren't there, and Yukie was already in bed. I wondered why I didn't realize from the very first that you wouldn't be at home. That came to me even before I heard a word of what she had to say. *I can do anything I want, or I needn't do anything. Given that this is a secret, it makes no difference.*

I remembered the long road on the hill and the vacant land right below your house. *The simple fact that I'm in such a place while the husband's not home permits me no excuses. At least in regards to Rumi. But I won't do anything.*

The two of us talked about when we were younger, and she repeated the bit about how she'd found it impossible to like me. Rumi once told you that in my presence. She then mentioned that you wanted to change jobs and work for a company that had more to offer.

"That's the important thing you wanted to talk about?"

"No, it's not."

"I don't understand."

"I'd like you to tell me whether or not I'm a normal woman. A normal woman doesn't cause a fuss, does she. She can patiently go on with things. And not be unfaithful. Is only our marriage different? There must be other couples like us. What do other women do in such a situation? I want you to tell me! You're the only person who can tell me about such things."

"You mean the problem is with Hiroshi?"

I stood up, leaned against the window, and lit a cigarette.

"The problem's with him. It's been no good for five years. But there are other marriages like that, aren't there. He says so himself. What's a woman to do when that happens? He won't tell me."

"That's a problem the couple has to deal with, and for a doctor, isn't it?"

"Of course it is. The problem is, I just can't do it. I wonder if I should cheat on him. I'm ashamed to admit it, but I'm half out of my mind."

"Wouldn't it be better if you talked this over with Hiroshi?"

"I need a man, in any case. That's what it comes down to. Besides, he says I'm an animal. Sometimes I think maybe I really am. Today, for example, I telephoned you in spite of my better judgment."

I felt I pretty much understood what Rumi was saying. The first thing that came to me was that this was indeed an important matter, and that your going out to play *gō* and Rumi's needing a man were both understandable. Were I Rumi, I would probably choose me. For a man.

"It doesn't mean I love you. Nothing of the sort. I've got to have a man. I need a man who won't lose his head. Understand? I'm not saying I love you. Accept that. That understood, there should be no problems."

"You're getting into a dangerous area, I'd say."

Rumi laughed for the first time.

"I'm sorry, but just talking about it makes me feel better. Am I a wicked woman?"

"At heart you doubtless possess many fine traits. But I have to be going now. I'll tell no one about our talk. It's as though it never happened."

"You're really leaving?"

"Yes."

"You're never coming back, are you."

I stood up without giving her an answer.

"But it's not fair to you to leave it at this, is it. Besides, Hiroshi does go out and leave you alone."

"It's not that way at all! *I* make it easy for him to go," she said, with some hesitation, her head down. "Hiroshi, on the other hand, drives me to near distraction, always watching me and meddling."

"Then he's watching right now?"

"He believes I won't leave Yukie at home and go out by myself."

"But he suspects something is going on here, doesn't he?"

"He's the kind of person who's sure that it could never happen in his own home. He'd never read my letters, you know."

Our fingers had brushed together as I closed the door. I pulled back my hand in confusion and dashed headlong down the sloping street your house commands.

I realize you're trying to become a "good man," as they say. You leave Rumi alone at home and go out and play gō, but that may be, as Rumi says, merely a trap she could potentially fall into. You've set it to measure Rumi's trustworthiness. Besides, by throwing yourself into the pastime of gō, which costs nothing, you're doing your best not to place a burden on your household finances. And in this way you're attempting to restrain Rumi. Which isn't to say it's my intention to reproach you for this, not in the least.

On the way home I found myself set upon by a feeling of indescribable light-headedness. I'd never been conscious of this feeling before. It was something foreign to me, and I dreaded meeting you on the way down the hill in this state of mind. When I got to the bottom of the hill and turned onto another street, I at last breathed a sigh of relief and reverted to my normal self again.

After my meeting with Rumi I kept thinking about the opportunity to make love to her. It was after she unburdened herself to me that I began to be pretty sure that I wouldn't do it. Rumi had divined the nature of that first stirring at the center of my being. It was there already as I came climbing up the hill, my shoes sounding on the road. This stirring within me had been the feeling I'd had that, by and large, a man is perfectly capable of making love to another man's wife if presented with the opportunity. I didn't have such a direct need in my relationship with Rumi.

Even now I sometimes feel an unbearable desire for my late wife. Putting aside the question of the women in our office, I've also engaged in sexual flights of fancy about women walking down the street, women I don't even know. About Rumi, however, I've never once indulged in such a fantasy.

Had I returned that night wrestling with my lust, I'd have gone looking for a prostitute or disposed of it some way, and that, paradoxically, would have even permitted me to lose myself in moralistic self-satisfaction. But I felt nothing. Even after I returned home I pictured in my mind the failure of your meaningless conjugal nights together. This excited me, and my setting off for your place while you were out was no longer a question for me of right or wrong. Well, perhaps it was wrong, but I'd at last be able to confront you. And I'd also be able to put a curse on my own flesh. Doesn't my stupidity make sense? But this insecure feeling, what do you make of it?

I'm obsessed with a compulsion for success, but it's entirely a compulsion for the techniques or wit to adroitly help you two succeed in this world, and what I must be sure to avoid is the feeling of being encircled by this wall of abstraction, your and Rumi's sexual problem. You and Rumi are being whipsawed by your physicality. And the mercilessness of this physicality—not your wife, not you—forces a confrontation upon me. My lending you the wit to conquer it is nothing more or less than my confronting this physicality. When you think about it, that's where my instability comes from. As I was reflecting on that, something that Rumi said came back to me.

"It doesn't mean I love you."

When I heard her say that I accepted it as something the straightforward Rumi would say, on top of which I felt I grasped the nature of her dilemma. Which also suggests that, love aside, should a man and a woman who've already experienced marriage embark on a relationship, they won't lose control of the situation. Rumi's infidelity has nothing to do with Rumi. It's merely a function of the physical.

That's all very well and good, but the undeniable fact is I've been provided a link to only your and Rumi's flesh.

The next day, when I saw that you looked absolutely no different than you always did, I immediately tried to put the whole thing down to delusion. But I was now in a comfortless situation. Things didn't go as well as they had the day before. I was distracted and unable to get myself in gear. I seemed to drift out of touch. I went back over what had happened and recalled every word Rumi uttered. Then I found that a scene—or rather, scenes—of Rumi beginning a liaison with a young man from the vegetable shop or the fishmonger's while you weren't home didn't easily present themselves to me. I would visualize, for example, the fish dealer who delivers to my house and Rumi suddenly embracing each other. But I couldn't visualize what might follow. Someone or something was canceling it out. All I was aware of was that the word adultery was spreading out over everything.

Why is it the fact that she's another man's wife so obliterates my desire and places upon my shoulders the gray burden of obligation?

For the next two days I exhausted myself in a battle with *the physical,* damnable and unseen. The result manifested itself in my ludicrous behavior. That night at the section party, first of all, I began by providing—to your astonishment—a complete account in the fullest detail of my sexual relations with my late wife. Next I bellowed at the boss for not sending me somewhere on business. I fell upon a woman from the office and demanded a kiss. Then, when you tried to stop me, I remember saying something like the following.

"You're just crazy about Rumi, aren't you. She loves you too. Next time around I'm gonna kiss her just like this, and you can watch!"

Finally, shouting, I demanded to be taken to see one of those loathsome lesbian or straight performances in the Asakusa district.

You and everyone else came to the unanimous conclusion I was so starved for sex since I'd lost my wife three years ago I'd lost all sense of proportion. To a man it was agreed I would be taken to a certain establishment in Asakusa and afforded consolation.

That was when you said: "Won't it make matters worse if we take him there and the guy gets all the more excited?"

I heard you refer to me as "the guy" and my drink-befuddled eyes lit up. I abruptly stood.

"Gentlemen!" I said, reverting to archaic diction, "Hiroshi well understands me."

"Right! Right you are!" someone responded, and we piled into a cab that would take us there. In the cab I let myself lean against you. Your rather thick-set body brushed against my cheek or shoulder every time the taxi swayed. Giving myself up to my drunkenness, I put my fingers under your collar, assuming a position which threatened further caresses.

"I'm a beast, so what do you expect? I hunger for it. I'm going to watch a beastly performance."

I was naturally playing Rumi's wifely role, but there was no reason you should've realized that. If anybody in the taxi wanted any less to watch "specialty exhibitionists" than you, it was me. Our reasons for feeling so differed, however.

The alcohol soon seemed to start having an effect on me, and I began caressing you in earnest, touching you all over. All the while you made no move to stop me.

Arriving at our destination, we waited mutely like a gang of thieves collared by the police on the first floor of a seamy establishment for the next performance to begin at last. Shivering in my increasing sobriety, I stole a quick look at your face in profile. You alone sat in formal posture, book open, studiously ignoring your surroundings. You appeared to have had a book on $g\bar{o}$ in your briefcase. Were you really unconcerned? Or was it a kind of pose?

Our group exchanged glances with the previous customers as they filed down the stairs. They were middle-aged men. All those guys have wives. Are their marriages going along okay? With what emotions will each of them be burdened as he returns home?

"Hey!" the section chief commanded, "we're here thanks to you, so you go up first!"

It wasn't my intention to lure you there. I simply was unable to suppress my own feelings and shouted that we must go there.

"And now we would like to begin." The performers, a man and a woman, commenced their presentation. The man's complexion suggested he was on the drug Philopon, but he had a well-developed organ. You read on in your book, briefly looking up from it, then returning your gaze to the page.

"How about that," marveled the boss. "It's like watching people work out. The rhythm they give to it is magnificent. Of course, without rhythm, it'd be an obscenity not worth the watching."

A good-natured smile played across your lips as the section chief spoke. Even here you sat formally on your heels.

"It would be as thoroughly annoying as my wife's workouts," you answered, then turned to me. "How about it? I'll bet you're going to your own show after this is over."

"Sure am."

At which point the couple bowed and bade us farewell, disappearing behind a sliding door.

I resolved at that moment to have my own performance with

Rumi. I'd then be able to think of myself as redeemed. I hadn't been excited by the show. Were I to go to your house while you were out, we would go from "and now we would like to begin" to the bow and farewell. The man would be none other than me. There'd be no spectators, but precisely because there were none, the tediousness of it would be much the same. If you're going to say in my presence "it would be as thoroughly annoying" as Rumi's exercising, then I've no alternative but to work out myself. Watch me. I'm on my way to my live-action show. With Rumi. Not out of love for Rumi. Hey, it's just friendship. And even if it's not friendship, I know it should be.

I'm just talking about assaulting Rumi's flesh, then keeping her quiet about it. I've always been oppressed by this damned physicality, and I've felt like I was floating in space, but I was now able to seize upon something meaningful, something I'd been waiting for.

Even here I gave no thought to Rumi's body (though you may find that hard to believe).

I'd apparently drunk myself sick. I felt chilled and queasy, so I wanted to be the first outside, and I went down the stairs and began putting my shoes on. And there before me in a windbreaker stood the male performer we had seen moments earlier.

"Thank you for coming. Here, would you like a shoehorn? Here you are."

He handed me a long yellow celluloid shoehorn. There was a long elastic cord attached to it, and when you let go of it, it sprang unfailingly back to its original pendant position. It was really a resourceful contrivance.

"This is," I commented, "a pleasant little bonus, eh what?"

"Yes, it's good to have it set up like that."

Continuing our singular exchange, we went outside, and as we did so it occurred to me that the live-act performer and the person taking on the role of guide were one and the same, and that this was singular, and that I was somehow being grievously insulted, and all the while my stomach was beginning to bother me something terrible, so that as I straddled a ditch it convulsed violently.

"Yorita's throwing up!"

You came around behind me and patted me on the back.

"A man gets self-indulgent when he's lonely. Don't overdo it.

How about staying at my place tonight? Rumiko'd be delighted to see you. We'll have her look after you. What do you say? Stay over. She's got too much time on her hands as it is."

From the way you talked it sounded as though you two were the ones who were self-indulgent.

"I'm gonna stay around here tonight."

I shook off your hand, got into a cab that had pulled up, and had the cabbie take me further into the Asakusa district, then circle back and take me to my sister's place.

The next day I stayed home from the office. I phoned and pleaded a headache—nothing serious, of course—saying I'd come in the following day, then, quite soon after you'd set off for work, I at last climbed the sloping street on which the morning sun shone so brightly.

I'd made up my mind the night before, yet when I considered my decision I was once again overwhelmed by an uncontrollable sense of the precarious. As I dialed the office number I thought: *and so now I have to go.*

3

As I went up the hill I thought: *You can still turn back, it's not too late.* You may not believe this, but at a time like that a hill has a certain ineffable power. Your hill is rather steep. If you climb it nonstop, it'll wind you. The simple act of climbing it gave me the feeling I was advancing in the face of undefined resistance of some sort.

The telephone and this sloping street had directed me toward the house from which you were absent. When I came to the vacant lot, which affords an unobstructed view, I caught sight of Rumi, your wife, moving about in the house. Rumi, as usual, was waiting, her ears straining for a football. And the owner of that footfall was yours truly.

What went through Rumi's mind when she saw see me standing before her, out of breath from coming up the hill (I who'd told her she certainly ought not to expect me)? I'd finally come and Rumi detected my approaching footsteps, but she listened more circumspectly than before. I'd come today to carry out something that I could've done two days ago. She certainly must've interpreted my letting a full two days go by as simple indecisiveness.

A few minutes later I'd at last say: "And now we would like to begin." Later, I'd bid her farewell. What else need I say to Rumi? Yet again, however, overwhelmed by the sensation that I was floating on air, I stood outside the door. I wish you could have seen me at that moment. I'd have liked you to have been there so that I could say to you: "What I'm going to do now means to you 'to have relations.' It's okay? It's okay, isn't it?"

I should've once again summoned you up as you were last night to break away from my floating sensation. By then, however, I'd already entered your house.

Rumi affected a casual air, causing me to wonder if this was the same person who'd earlier made her needs known so baldly. Yukie was at school. Only Rumi and I in your home. I looked at the view outside as I wordlessly drank my tea. From your house, which seems almost to sit atop a scenic promontory, I could see some housewives doing their laundry and several other housewives scurrying about at the bell of the garbage truck. The nightmarish sexual exhibition of the previous night passed before my eyes. This homey scene of housewives, having coupled with their husbands last night. Well, to begin with, what are we to make of life or of the things people do? In the room books on economics that I rather doubt you look at anymore. Their slipcases are yellowed by the sun. On another bookshelf, children's picturebooks and guides to dressmaking. Someone in celebration of your wedding doubtless presented you two with the coffee cup I drank from. I'll bet Rumi at least discussed the pattern of this almost new carpet with you, even if you didn't choose it yourself. The family photo album you showed me when I was first invited here. The history of a family. A camera hanging from a post.

What was Rumi doing? She was now peeling an apple. You're wondering what the hell I'm talking about. At that moment, I wished I might disappear from the face of the earth. I couldn't bear to have her peeling the apple.

Rumi's odor was in the air. I heard her quietly approaching.

"Nice day today," I say without looking at her, then only, "No work today; I've got a hangover."

"Hiroshi was late getting home too. I asked him about it, but he wouldn't tell me anything. You had a good time?"

Rumi didn't look up either. A comfortable Old Rose brand

sweater. The sun shone on the nape of her neck. A slender neck. Her hair with its carefully set waves. An honest-to-God housewife.

No sooner did I realize we were both reaching out for the same slice of apple than her fingertips brushed ever so slightly against mine. The instant her fingers took hold of my fingers she began shaking convulsively. As I had feared two days before, this imperceptible contact of our fingertips suggested the path we were to follow.

Rumi held tightly onto my fingertips and began to weep. What was it that was making her cry like this? Was it fear? Was it sorrow? Should I presently commence with the task before me? The absurd, imperative moment, the one that I had most feared, was finally at hand. I took Rumi into my arms. Her shaking grew all the more intense.

"Rumi, let's stop. Your trembling like this, that in itself tells you it's not right."

"I won't let you leave!" Rumi screamed, shaking her head. "You can't go! I refuse to let you leave!"

"But Rumi," I protested more than once, "you're terrified."

If she's terrified, shouldn't it be only of Hiroshi, the man? What else is there to be afraid of? Isn't that what I've continued to dwell on? If *Rumi* herself has not come up with a solution to her problem, we would have to stop. I moved to take my hands from her back. Now, I thought, I should flee down the hill like a shot. Rumi seized my retreating hands with a fierce, vicelike grip.

"It's not my husband I fear. That's not it. Do you understand? Do you? It's me I'm afraid of. My body. I myself don't know what kind of mess I might stir up, what I might do. I'm so ashamed. In the state I'm in, it's you that terrifies me. Do you understand? I don't want to let you go, and I'm in fear of you. It's more than I can bear, it really is."

It was then, Hiroshi, that I performed an act that—at least in terms of its mechanics—was indistinguishable from "becoming intimate," as they say, but as I did so I struggled against her flesh. That, I think, you especially would understand. And the smell that came to me was the odor of death. Isn't that how you look at Rumi?

I watched my wife die. I've so far seen seven or eight deaths in my family. The convulsions at the moment of death and the collapse that then sets upon them. They're a manifestation of the body fighting frantically against the violence within itself. There is, of course, agony in dying. Yet Rumi's ecstasy is not unrelated to that agony. The anguish of a black beast, of a black flame. It was my intention to triumph over it, and so I accomplished my ridiculous task. I'd completed my first mission. And what do you think, my friend? Never before had I felt less desire than I had at that time. Will you again think me a liar if I tell you I experienced only a sense of pathos and farce?

All along I'd felt my relationship with Rumi farcical, but in a different sense. It was the absurdity of this relationship wherein I become a tool, having discussed becoming a tool and finding myself agreeing to the thrust of that discussion and therefore at my wit's end. There remained yet a ray of hope for me. It lay, I suspect, in being able, through my act, to somehow steal from the two of you, man and wife, the savor of adultery. It would have been in my feeling desire. My hope of being overwhelmed by it, losing myself to pure lust at the sight of Rumi's body. Were that to happen, I could've become a scoundrel in your eyes. But nothing of the sort happened. I thought of you, but this time I felt no desire for Rumi's body. I felt nothing, and yet did what was to be done. What do you make of that?

I've become uneasy about what Rumi might think of me. Surely she's unaware that I feel nothing for her. If she were not, she would feel deeply ashamed for her body. I thought about it and decided to feign ignorance.

"Tell me, I wonder if I really am the animal Hiroshi says I am."

"Animal? Animals are more wholesome than humans. Humans are more difficult to deal with than animals."

"What do you mean by that? You're saying I'm already a burden to you?"

"I don't know anything about that."

"You know nothing about that? What do you mean you know nothing about it? You're playing games with me. You are, aren't you. I don't want your sympathy."

Exasperated at this point in our conversation, I stood up.

"Why can't you simply think of yourself as a tool?"

"You're the tool! And you'll come back again, tool that you are. And you must make me your tool, while you're about it. I know you will. It's just that you're suppressing it. I know how that is.

Listen, do you follow me? If you don't come back, I'll look elsewhere, or I'll kill myself."

You're dead each time, aren't you, Hiroshi.

I get nothing out of it. I merely feel I've seen death. Perhaps a woman like Rumi is meant to project that image, then die.

"We have to keep this a secret from Hiroshi, it's our responsibility. That's all I ask, so promise me that."

"That's why I wanted you to come *here*," Rumi responded. "We've nothing to worry about if you come here. If he finds us together, I can talk our way out of it, no problem. Yukie will never tell on you. She hates her daddy."

Eyes closed, Rumi threw herself at me. I caught her like one might catch hold of a rotted tree that falls and kissed her on the forehead. Rumi's fit of shaking set in again. I mouthed sweet nothings, but felt only loathing. There was no other tack for me to take, for I also held my hate-filled heart in contempt.

"Yukie will be back any minute," I said.

"We'll hear her footsteps when she comes. You coward."

Once again I mechanically satisfied Rumi's needs, after which she made me tell her when I would visit her again.

On the way down the hill I caught sight of little Yukie. When she saw me—she had seen me only once before—she called out and came running up. I realized for the first time that her face was the spitting image of yours. I looked intently, too intently, into her eyes.

"What's the matter?" she asked. I told her everything was fine and, pointing toward the house, added that she should hurry home. I don't know why I told her to hurry home. Perhaps it was the desire to hurl life into the midst of death.

4

Damn you, Hiroshi, I'll bet you've seen death in Rumi for some time now. And though I may have sated this death, in the end, you and I are alike in being mere observers of it.

I descended the hill, and as I stood on the corner, I heard the noon sirens sounding over the city. They rose irritatingly, and after long, self-indulgent wails, were gone. Curiously, I recalled your going out to lunch. I haven't gone much with you. That's because you need to have someone who can talk *gō* with you even at lunch. So

I merely watched you go out to lunch, but I've felt like I was walking along with you.

I sensed for the first time just how difficult it is to keep a secret from you. Because, once again, I've been overwhelmed by uneasiness. I stood on the streetcorner not knowing where to go. I had no place to go. If I'm away from Rumi, it seems you can be my only destination. So I went to the office in the afternoon. When you saw me, you seemed heedless of both last night and of the fact that I'd not been at work that morning, and came out with your customary: "You learn gō, too. I'll show you the ropes." What do you think my thoughts were then?

Well, it looks as though I have to continue my relationship with Rumi.

When I saw that your cigarettes were Hikaris, I left your place and immediately went out to buy a pack for myself, resolving to get into the habit of buying them. On my way out I noticed one of your shoes was sticking out from your shoe shelf by the front door, so I pushed it back in. You're the kind who doesn't wear his soles down unevenly, aren't you.

The next time I set off for your house, someone was coming toward me down the hill. I was startled, for I heard your heavy footfall. I went back to the corner, then turned left and ducked behind a fence to let you go by. That afternoon I'd had a call at the office from Rumi, and we had agreed on a time, but you apparently left the house thirty minutes behind schedule. You stood on the corner, pausing for a moment, stuck your hands into your pockets, and looked back at the house as though you'd forgotten something. Then you went back to the house.

As you were coming back to the corner, I was still waiting behind the fence, and I lit a Hikari. After I left Rumi I'd been giving some thought to these three days. The reason was that time was passing and I'd not been able to come up with anything. Rumi in that house. A dying Rumi's voice, in near agony from her suffering. You at the office. In my fantasies you steadfastly refuse to oblige me by staying home. Only Rumi is in that house. I'm nowhere to be found. You, at the office, overlap with me. The reason I was surprised when I saw you was because you actually were coming down the hill.

I immediately realized from the bulge in your overcoat pocket that it was your *Introduction to Gō* that you had forgotten. You threw down your cigarette in front of me at the corner, urinated a rather good spell, making me wish I could vanish into thin air. You had obviously drunk some beer. When you finished, you spit for good measure. You then looked up at the sky. It was a cold night, yet a halo of mist surrounded the moon. I also looked up from behind the fence and read the sky. Why had you looked up at the sky? A man who looks skyward is a man without cares. Without doubt you're completely up front about Rumi. Or perhaps Rumi had now stopped making her appeals to you and you could even feel a sense of liberation at Rumi's transformation. You probably didn't sense that it was pain, assuming that you even wondered what had happened to her.

Never having had such an experience with my wife and not being a writer of fiction, I've come to the conclusion your feelings for her are as utterly insubstantial as a handful of cloud.

I paused to catch my breath halfway up the slope. I couldn't feel the surge of energy climbing here I'd felt the other day.

That woman must live. Yet she must die.

In my mind I considered something I could not myself really comprehend. I eventually began walking again. It was, after all, a street.

I caught sight of Rumi running down the hill toward me. There was no need for any more thinking.

"I thought it was you a little while ago, but it was Hiroshi coming back. I did get quite a jolt in spite of myself. I was sure you'd run into him at the top of the hill. But if you do run into him anytime, come up together, okay?"

"I was at the bottom of the hill."

"Thank God!" she said, hugging me and sighing deeply. "You know, dear, it's been a long wait. Very long. Forgive me, but I've fallen for you. I was contemptuous of falling in love. The way I feel now is what they mean by love. You know, I really despised myself. I did."

Rumi suddenly realized we were at the top of the hill. She cut herself short, took my hand, and pulled me along at a quick pace.

"Do you understand what I'm saying? Just a little while ago I toyed with the idea of throwing myself off that bluff. You know, it's weird, this feeling of being in love. I was better off before. I've been dying a slow death these last three days. I didn't give a thought to men up to three days ago. And suddenly feeling like this, it's strange, very strange. I'm not sure of myself again. I seem to be someone who's ruled by her emotions."

That, too, was Death's doing, making her proclaim her love for me.

"You're going to catch cold, Rumi."

"This feeling, what can I do to control it? You must know. You're the one who's made me feel this way. You know how, don't you?"

Rumi suddenly took a step back.

"Wait a minute!"

I could see in the moonlight her mouth in a wry twist.

"Wait a minute! It's all one way, isn't it! What sort of expression is that, that sullen face? I retract everything I said just now. It was all a lie."

She took my arm. "Well, step inside."

"In I go to do my job, eh," I said with intentional spite, though my own words disgusted me. Rumi's response was even more spiteful.

"That's right! You can think of me as a streetwalking whore, can't you!"

What is there to say? The Festival of Death is to begin.

The smell of hair drew near.

Hiroshi, for the third time I stood face-to-face with my own wrongdoing. Thus it was that at the time this curious connection between Rumi and me approached nearest to the phrase "have relations with," I would sense the instant of its severance.

I wonder, Hiroshi, if this sounds like another attempt to talk my way out of something? Had Rumi been my wife all along I wouldn't have been confronted with this sort of wrongdoing, each on the heels of the other. A person is capable of feeling desire even if he doesn't love his partner. And he can fall in love simply by feeling desire. One can even love death. For me, however, you were there from the very beginning. I have no idea why your existence, on the contrary, did not enhance my passion. What am I?

5

I'm going to tear up this letter and throw it away. Ultimately, this letter is proof I continue to feel a closeness for Hiroshi. And I seem to have written it as quite possibly a precaution against revealing the secret to him at some point. The truth is, I may well just sidle up to him some day and whisper it into his ear at the worst possible moment. And that, indeed, in the last analysis, would seem to be the true nature of the almost pleasurable feeling of dizziness that has again plagued me all day today. The reason I gave Rumi's hand a tight squeeze after she had squeezed my hand was, I think, because I wanted to verify to myself the existence of this precaution.

There is no doubt whatsoever that once again tonight I'll climb up that sloping street, my mind robbed of its self-possession. I even suspect Hiroshi—laughing on the sloping street—will suddenly put his hand on my shoulder.

"Go for it!" he'll say, and I'll probably send him flying. Because it's impossible to tell him in the space of one second everything I've written in the letter. If anybody's going to tell him, he's going to hear it from me, no one else. I won't hold anything back. On the contrary, even now I've got half a mind to summon him there when we're together. "The fact is," I'd tell him, "I've fallen in love with Rumi." For I'd have to do it this time with words, words from within the hurricane born of her possession by the Angel of Death, in order to placate it. She'd like to think I come running up the hill burning with love for her.

The only way for me to fall in love with Rumi is for her to despair of my love, not for her to throw herself off a cliff. I propose the following scenario: Hiroshi, returning from an evening of *gō*, finds us in the act and is driven to despair, and Rumi is driven to despair, thereupon there would be nothing left for me but despair. And I think now I want to fall in love with Rumi. Yet if even *that* does not drive Hiroshi to despair, won't I, in the end, be a mere performer?

TWO POEMS

FORREST GANDER

CLOSE UP: DINER, BREAKFAST

The door opens at five.
    What would we like this a.m.,
                      the waitress asks,
though I am alone.

            Seated under a sign
       for infrared broiling, I see
the coming loose of her face at its corners.
But she does not affirm herself,
               she prevails.

            The character you are looking for
is not in this poem, was here earlier
and had to leave. Suddenly,

        I recognize my own face
   in the waitress's eye.

Like the character who is absent
I know all that I can know:

                        tender undertones in winter, porchsteps
            which need pointing,
                    frogs
                            signifying coolness,
            the weary, random seasons in hell.

TIME LAPSE: DINER, EVENING
                                            Inside, the air
                                            supersaturates
as on a morning this summer
when someone wore dog shit to church
under his shoe but didn't notice
and the congregation breathed through their mouths
like trout, their attention gone,
sweating, until he crossed his knees
during the sermon, smeared
a swatch across his cuff,
and exited.
                                All the fish you can eat: $5.95.
        I'm over
            from the slender, one-armed mechanic. Each person
                involved but isolate
like deer beyond the parking lot, grazing on sunset.
                                    On the next red stool
someone
                insinuates her knife
behind gill rakers along the brain case,
        down the dorsal shield
                    and across the tail, so
pale meat unzips from bone.
                    Deft touch,
I say, though she is eating,
                and so continue.

SIX POEMS

CAROLYN STOLOFF

IN THIS QUADRANT OF THE MOON

what is your port? your given name?
are you obsessed with Alaska
or Egypt

for what do you hunger? salt
mouths   coins?
are you a Perseus
or a stone-face

are you luminous
do you live in a red house? a canyon?

I dismount   now you
speak   give evidence

have you filled your saddlebags
with alfalfa   or gold

do you read the text of daylight
on the given field
or dowse for a darker domain

in this quadrant of the moon
I am the Empress of Russia

the King of Castille
the last of the ancient Papuans

scattering bones on the plain
I divine you   reader
fellow lunatic
you are my alibi

running parallel, you too swallow pain
spooned out by the clock's hands
catch light   hide fear

we sing the same song:
who are you? I am here

FEBRUARY 8

poles are down   at the city's fringe
boxcars sleep under ermine   wind
shakes huge sheets

an arctic fox
pads invisibly through the street
tongue like a ship's flag

at sea, a fog horn complains
it's too far between hearts

that's why snow mice
huddle on sills
frozen milk slides
down dim deserted chimneys

at each corner a statue
clutches a cup of ice

it's too far to go
I miss your face on my pillow

on my bed a pale lizard lies
reading the ceiling's
simple page

a pipe raps
no . . . it's the blind widow
making her way

AGAINST THE SHINING WIND

Brooklyn Botanic Gardens

idling beneath trees, I close my palm
against the shining wind
that would take

the grass print from my cheek

spring's sometime will never come
I think   skin deep in shells
in calcium petals

in the white tree top
waist deep   a girl in red

her hands' open doors
frame darkness   her mouth
calls down the corridor

between cherries blossoming

she calls her horse

still far away   calling
she feels a thundering not hoofs

her blood gallops

I had forgotten how the wind
knocks blossoms down

CLOSE CALL

in a room papered with war the telephone rings
a circle of numbers drifts toward the ceiling

I close my hand on the black tongue
babies with open mouths ooze from the receiver

a voice from the grave staggers   your voice

a paper rose sprouts from the receiver
the smell of mould   feet   urine

an ocean slipping

from the receiver's sieve something green
delicate as lettuce—

something clean—falls to my lap
what the sea left on your brain

its whiteness love

SUNDAY

a hand
wards off the wolf smile
at the skylight—

the heaped day
asleep on the passionless couch
growls through dreams
scented with basil

we're both sick of talk
and the sink overflowing

I plunge my arm in
for the stopper

over the square—a white sky
combed by grey elms
bare, except for a tilting nest—
its inner curve worked smooth
some spring ago

hour-beads
hang heavy from my neck
as your spine sinks into cushions

have we given up
dancing for good?
is Little Red Riding Hood's hood
pulled tight at the throat,
as we're drawn
down a gullet,

still seeking partners
still listening for a duet

ON THE LONG FALL DOWN THE VAULT STEPS STOPPED

Turned by the *thump thump*, struck
by the sight of her, supine, by her silence—
she who rained me out once,
she whom I mother now, puddled
on the leveling floor with legs angled up
wide-eyed, addled, not yet horrified
at having been urged earthward,
at having let go—brittle column collapsed. . . .

Oh rise, rise jessant
as a spring, rise as a sprout from the cross-cut
the axed heartwood, the stump
left from what planked my child-bureau brushed
with spring green, rosebudded with sealing wax,
in whose stuck drawers we pose, caught
in emulsion, curtseying in Kate
Greenaway calico dresses among balsam pillows,
among dried flowers, matchbooks, petticoats.

Rise as a shoot from the root-stock of rosebush,
winter indignity over. Grow, branch, bud
to be plucked again, pink satin indulgences
you adored, for the grey chiffon
loosened in waltzing, in being turned.

Stop. Don't be blown
open. Don't petal down, don't.
What we in living did we must undo—
our nots, knotted by us (if not by us, who?)
as we shrink, as we trip into fall.

THE JAZZ CLASS

WILLIAM WEBB

As Karen Ingram walked home from her Tuesday afternoon jazz class, thoughts of the upcoming recital so absorbed her that she found herself almost underneath the scaffold before she noticed it. She hadn't had any particular reason to crank her head back and stare up at the gray sky, and so she hadn't, whereas any chance of her spotting the two workmen predicated just such an action (cranking, staring, etc.), since the narrowness of the courtyard, its perpendicularity to the street, and the leaves and branches of an expansive ginko tree all combined to obscure that side of the building from a distance. As a result, she did not see the men or expect them to be there. In fact, her mind was on other things.[1]

The scaffold dangled at a precipitous angle from four stories in the air, and two men worked feverishly at the ropes to straighten it. Karen figured the men to be painters; they wore white overalls and painters' caps. The possibility of their falling did not occur to her at first. The inherent danger of working on a scaffold seemed to preclude such an accident, just as the perilous nature of tightrope walking or trapeze artistry disallows a mistake in the eyes of an observer. It's an act, after all—one that's been practiced again and again. The performers balance on chairs and on bicycles; their bodies somersault through the air to the crashes of cymbals. But these tricks are merely feats of skill, and somehow their very ac-

[1] The analysis of Ms. Ingram's internal state has been rigorously based on facts and testimony. Footnotes throughout the text will provide the concerned reader with an objective basis for interpretation.

complishment makes them unreal. Besides, the men on the scaffold brought to Karen's mind *The Guinness Book of World Records,* in which a window washer is reported to have been struck by lightning seven times. If he could live through multiple electrocutions, surely persons of a like temperament could withstand a fall of some four stories—these two painters, for instance. It should mean nothing to them.[2]

Before she happened on this odd situation, she had been deciding whether or not to let Ann Henry perform with her and two other girls in a jazz routine. The dance instructor, Mrs. Downs, had suggested the girls work on a "more advanced piece" and perform it on their own—Ann Henry included. They would have to practice outside of class; it would mean some extra work. But Mrs. Downs thought they were ready, and she would help them as much as they needed it. Her eyebrows inclined as if to say: how about it? (The expression had made her look almost French.)[3]

At first the girls barely nodded, rocking onto their toes and holding their hands behind their backs. Then they nodded yes more surely and smiled at each other. "Oh," said Mrs. Downs, mockconspiring with them and talking to them sideways, "one other thing. We'll have to get you some costumes that will really show off your legs." At which the girls let out titters and giggles that sounded something like air being expelled from the pinched neck of a balloon.[4]

Greg Wilkenson provides an eyewitness account of the accident: "I had just finished practicing my cornet. I hate that instrument. I mean, I hate practicing it. Other things about it—the shiny brass, the smell of the valve oil, and the feel of the red velvet-lined case—I like. Anyway, I was about ready for some chocolate milk—which

[2] In point of fact, Ray C. Sullivan—a park ranger rather than a window washer—is the only man to have lived after reportedly being struck by lightning seven times. He died by taking his own life in 1973 after suffering from a broken heart. Norris McWhirter, ed., *Guinness Book of World Records* (New York: Bantam Books, 1986), p. 451.
[3] Lillian Bragdon, *The Land and People of France* (New York: J. B. Lippincott Co., 1960), p. 30.
[4] Based on Ann Henry's testimony and the opinion of Rusty Hamhold, high school physics teacher.

you can't drink before you play, because it gums up your mouth. I had just poured a tall glass of chocolate milk, when I happened to look out the window and see the two men working on the opposite side of the courtyard about ready to fall and Karen Ingram walking by on the sidewalk. She didn't see me."[5]

The two men on the scaffold seemed to be in a predicament. Their efforts to right the platform had only tipped it farther, and the pulley system reduced their frantic counterefforts to a slim, inch-by-inch improvement. At any moment they might plunge from the scaffold dragging brushes, cans, and scrapers with them. Karen was glad not to be in their shoes.

She became concerned for herself only when a man in a Cadillac began shouting obscenities. "Will you at least let me get out of here!" he spouted out amidst four-letter words. Karen backed up a few steps. Perhaps the workmen endangered her too, she didn't know. The man in the Cadillac wore rose-colored glasses that made him look effeminate. His hair had been tinted blue-white, and his face looked oddly wrinkled, as if the wrinkles were pasted on. He yanked the steering wheel all the way to the right and nudged into the back bumper of the car parked in front of him. Just then, a bus pulled alongside and blocked his escape. He began to honk and wave his fist at the bus driver.

Karen's tongue made a deprecating "tsk" from between her teeth. She just wanted to get home. Her jazz class had not gone well, and

[5] Greg knew Karen in an incidental way from school. Before classes started, when everyone milled in their respective groups around the circular upstairs corridor, he would see her walking with a varsity football player or flipping back her hair and laughing next to some girlfriend. He was a Brain, and she was Unattainable. Still, he often walked by the open door to Mr. Hamhold's physics class just to see her sitting there. For some reason, she usually sat with her legs spread apart. When Greg walked by, Mr. Hamhold would be addressing some solemn subject, like inertia or the unapproachable speed of light, his voice resonating calculus and algebra and the chalk nipping at the blackboard. Dazed, in a slouched-down grande plié with heels pressed together and thighs at a ninety-five degrees, she might be half-staring out the door and watching the traffic. Greg's mind would flash to take a picture of this almost too-wonderful girl-woman, Karen Ingram, whose triangular geometry and cool smoothness made him delirious. His mind's eye undressed her, touched her, and dwelled on every nubile curve.

she was tired and hungry. She wondered what her mother had made for dinner. Not chicken again, she hoped.

Ann Henry presented a problem. Why didn't the girl just lose some weight? She didn't seem to understand how awful the routine would look if she performed alongside Lisa, Debby, and Karen. Not that they didn't like Ann—they did! But this was a performance, and Ann would ruin it for them. All the audience needed to see was dumpy Ann Henry leaping around in tights, and they'd break out laughing. They'd laugh the girls right off the stage. Imagine how crushed Ann would feel then! Maybe she'd kill herself. Anyway, if the other girls excluded her, it was for Ann's own good.

Of course, no one had said, "Ann, you can't practice with us." For one thing, Mrs. Downs thought Ann would fit right in. Nevertheless, when the girls arranged to practice, Ann wasn't invited. The other girls knew why well enough. Debby had said, "Maybe we should let Ann know." And Lisa had said, "Well, I'll tell her if I see her." Naturally Ann never arrived when the girls got together in Debby Calarco's basement. Lisa Newton shrugged and said, "Well, I didn't tell her. I never ran into her." "Maybe next time," Karen had said, knowing full well Ann wouldn't be there next time either.

But then there were the costumes Mrs. Downs had ordered. The recital had crept up to two weeks away, and she wanted to make sure that everyone had one. Everyone did—Ann Henry included.

Oh, that was a scene. When Karen saw all the eggshell-textured boxes, she felt as if her younger brother had kicked her in the stomach. Mrs. Downs clapped her hands and ushered the girls to the center of the room. They could look at their costumes and try them on at home, since the class wasn't over. (It had just begun.) She handed out mimeographed sheets that would explain to their mothers how to hem up the fabric and where to sew on the trim. Then, while the other girls set down their costumes along the perimeter of the dance floor, Ann Henry brought hers up to the front as if she had a question on a test.

"I can't wear this," she said.

"Why not?" said Mrs. Downs. "You haven't even tried it on yet."

"I know. I just can't."

Without another word, she put on her snorkel coat and left. There were clouds in her eyes.

"Karen, Debby, and Lisa—I'd like to see you after class," said Mrs. Downs.

That class had been torture, and afterward Mrs. Downs gave the girls a good talking to. What were you thinking of? she kept saying. She reduced Lisa to tears. Karen felt like crying too, but she didn't feel as bad as she had when she watched Ann zip up her coat until just a saucer-sized hole was left for her face and then walk through the door outside, where she would have to wait another hour for her mother.[6]

The men on the scaffold seemed ready to crash down at any moment, yet they hung on, neither recovering a safer position nor slipping into a riskier one. Their pulley system must have jammed somehow. Karen had been standing there for about half a minute, but it seemed like much longer. If they were going to fall, she could spend all day waiting for them. She wished they would fall and get it over with.

She realized that the blue-haired man in the Cadillac had nothing to worry about. The painters were too far away to affect him. For that matter, she judged the sidewalk to be in safe territory. Why hadn't she seen so in the first place? It seemed obvious to her now. The blue-haired man had tricked her into being afraid for no good reason.

She gathered her things and started forward. She was carrying boxes for two different dance outfits and a knapsack overstuffed with books, but despite these encumbrances, she took light steps as if she were dancing. Soon she *was* dancing: her feet skipped along in rhythm; her body swayed and turned as if to music. Leaning back, she kicked high into the air. If I could only see myself! she thought, smirking. When she noticed that the blue-haired man was staring at her, his total surprise made her laugh out loud. She glanced up at the men on the scaffold, who smiled back at her. She never imagined that she might be distracting them.

At that moment, Ann Henry was solacing herself over a hot fudge sundae. She'd already relayed her situation through sobs and tears to her mother, and together they'd wondered at "the meanness of

---

[6] Actually, according to Ann Henry, she only waited twenty minutes or so.

some people." Convinced now that she'd never really wanted to perform "something more advanced," Ann's only concern was for the E period physics class tomorrow, which she happened to share with Karen Ingram. More than likely, Karen would pretend that she had nothing to do with Ann's being left out. That falseness hurt the most. Karen would never say, "You're just too fat," even if it was true. Was it true? Mrs. Downs didn't seem to think so, but maybe she was just being *nice*. If he suspected that she was upset, Mr. Hamhold might attempt to console Ann as well. Lately he'd been giving her the creeps the way he leaned too close and always wanted to help her. Once he'd even touched her knee and said, "You're different from the other girls." She could picture him looking crowlike with coke-bottle glasses while he waited for his explanation of vectors to sink in. He could be very kind, but sometimes he sat so close that he breathed in her face. She hated that.[7]

Greg Wilkenson was watching Karen as she performed her jazz routine on the sidewalk. Without turning away he sipped his chocolate milk. Although his sister took jazz and ballet, the aesthetics of dance confounded him. Why were these certain steps and positions to be used rather than some others? He liked watching cheerleaders better, because the girls didn't wear pink tights on their legs.

He'd been to several dance recitals, so he knew how they went. The high school parking lot would jam full of cars, most of them

---

[7] Teenagers routinely misinterpret communication as being sexual when nothing sexual was intended. They've got sex on the brain. Moreover, adolescents—girls especially, if you'll forgive my chauvinism—often hold an inflated view of themselves, so that another person's appearance and comments must be intimately related to themselves. The world seems to revolve around the "I." It doesn't—I can prove that to you.

There's nothing wrong with my encouraging a fine student. Teachers don't get enough of the good ones. But as for student-teacher relations, I maintain a professional distance. Of course, a complimentary pat now and then may reassure a student of a teacher's humanness. And classroom discipline often involves some physical contact—it's unavoidable.

I'll never forget the time I substituted for a girl's school in Waynesborough. What brats! Finally, as a last resort, I brought the worst behaved of the bunch into a back room, bent her over my lap, hiked up her skirt, pulled down her panties, and spanked her bare bottom with a ruler. I experienced some degree of pleasure as a result, I'll admit, but my actions were directed by pure necessity.

bigger and more expensive than his parents' VW bus. The men stepped out in coats and ties and led their wives and daughters up the stairs. It was like church, except that a lot more people went.

Inside, parents and siblings milled about the lobby and stared at the recessed lighting while they waited for the show to begin. Often they arrived quite early, since the girls needed time to change into costumes and to put on makeup. Acquaintances struck up polite conversations in which the women used high-pitched voices and agreed on many points. The men stood around awkwardly, asking each other what they did for a living. Everywhere there was the swish of clothing, the clip-clop of high heels, and the smell of cigarette smoke. After the husbands had fidgeted for a sufficient length of time, they accepted programs from a girl in an orange taffeta dress and found seats for their wives in the auditorium.

Meanwhile, Greg knew, arcane rituals were being performed in the jam-packed girl's bathroom, for which the girls had to strip off all their clothes to change, since underwear would show up beneath their costumes. Even six- and seven-year-olds were made to wear make-up, and the make-up in combination with the hot stage lights made all the girls sweat like pigs.

Karen felt as if she were dancing in a dream; every step and turn seemed inevitable. Not intending to, she'd begun her jazz routine. She knew she was right: the world was made of appearances; they spun around her. Lisa and Debby made good friends because they were pretty. Together with Karen they'd perform in the recital just as they'd practiced for it, without Ann Henry. What could Mrs. Downs do? She couldn't stop them. Besides, she'd want them to perform—her best and her most beautiful. There'd been a misunderstanding, but now there wasn't time for Ann to learn a whole routine. No way would she dance with the rest of them: she was just too fat.

Someone shouted out—one of the men on the scaffold. Waving his arms like propellers, he teetered on the edge of space. Then, at the last possible moment, he grabbed for a rope and held on. He saved himself from falling, but the platform bobbed and shook and dislodged a full, gallon can of paint. It skidded down the scaffold's steep incline and gathered momentum along a horizontal axis; it tipped off the end in a graceful parabola. Deflecting off the upper branches of the ginko tree, the paint can sailed out over the sidewalk and crashed down on Karen's head.

She fell face-forward onto the concrete. Her boxes and books lay scattered around her. The blue-haired man leapt from his Cadillac, but what could he do? Not wanting to touch her body, he picked up a box that had spilled open and pulled out a leotard covered with red sequins. It sparkled in his hands.

SIX PROSE-POEMS

BARRY SILESKY

GROUND ZERO

For weeks the rumors have been spreading and we feel the excitement. Someone's ringing a bell, someone's shouting, the woman next door is calling her cats, scanning the windows. There must be a man who will listen, split a beer, explain the heat, the prices, the natives across the street gathered on porches, mumbling their guttural tongue. Is that plane the first sign? The child crying next door? Or something we can't hear? Downtown the store aisles are jammed for the sales, the groceries are selling out. Maybe the ballgame won't be rained out. Tonight we'll try that new restaurant. With the tv off it's hard to know for sure; perhaps we made a mistake and the train is on time, the bill due yesterday. Her name was L——, but it's too late now. At the cab stand they've started a pool on the location & time. The Arab shouts it's his choice first, the Korean raises his fist, the white boy snickers and spits. The bystanders laugh and try to move out of the way. They know this is it.

SWAMP WITH GOLDFINCH

The black bars so stark against the bright wings, the lone dead branches rising from the grass. The sudden shock amid the vast expanse of green and blue surrounds our attention: forget the revolution, the scandal at the Capital. The natives surrounding the market? Let them die without us. The one victory we knew is enough for hope. It really happened, I swear it. The screams of the executions filling the stadium after the final defeat can't be ours. Everyone knows the papers lie. The beautiful gold bird they'll never see stuns the day with its music; when he disappears, we're just grateful he's seen us. Tomorrow, next week, the company begins arriving and before we know, the children and their noisy complaints we want to escape have done it themselves. Of course they'll be back, at least one of them, relative or friend, another version, stronger, the difference the years have made subtle but crucial: they're ready to lead us. For years we've described the country they're born to and we half believe it. We wish them luck. Perhaps they'll save something despite us.

AFTERNOON OASIS

The tips of the aster leaves rust, the first sign of the disease, but it's easy to overlook. Just an experiment; even when it dies, there's so much left: day lilies' orange, azalea, dogwood, beans and tomatoes so lush, even if the air makes them too risky to eat. At least it's cooler, and we can enjoy our square of yard. With plumbing to fix, appointments at 11 and 1, there so much to make sense of the day. Have another cup of coffee. But your legs are shaking— really, everyone needs a vacation, it won't last forever. Meanwhile, listen to the dogs howl, the horns blare at something the hedge blocks completely. You've lived here so long it's music. And the stench the breeze wafts from the alley? Such comfort to know we're not alone. When the siren wails again, we hold our breath in anticipation. Those gunshots are so exciting.

PRIME TIME

"How do you comport yourself with Jack the Ripper?" —G. Burns

Did you hear it? I have to look back: one eye, the gleam in black spilling out of its frame: a museum piece? Safe that way, but then he's in the mirror, drinking my brandy, and all I can think is sleep. It comes finally, but in patches; we've assembled the axe, the hammer, the knife next to the bed, as the knock rattles the window. Carefully, with aching slowness, one step on the carpet, another, tense in the still hall: anything but to wake that breathing. The train hums in the distance, it must be coming, but never soon enough. The wimper spills under the door, the stomach turns in the throat. Armor forged, mask donned, the years ravel out the window. I promise help, wish him luck, buy him a drink: have the tv, the stereo; here, let me help you with those packages, really, we all want the same thing.

Except, of course, he doesn't: leering out of the stairwell, it's another language completely. Put him away quick. But the cops don't believe me and the prisons are bursting. Why else is he here? Please, a last drink to deaden the skin.

AGAIN THE HERO

doesn't come. Sorry, too cold. We sympathize, really, but you understand since the car crash. He'll be better next week. Still, I want to prepare you: it'll never be the same. Not a limp, no permanent wheeze, and that graying shock spilling down the temples is impressive: a sign we've earned these complaints, a new car, the rich expanse of lawn. When we can't afford the nursemaid for the babies, we've finally arrived: the gray day the boat docks is a holiday; the bright young man arrives with his camera, and we sign the checks with delight. All the years of confusion finally passed, we've mastered the language, even if the inflection gives away the past. Still, it's what makes this country great: chicken on sale at the grocery, the neighborhood aisles an invitation we can't turn down. But hey, I've got an idea, let's order in tonight, call up the neighbors. The attack's receding now, maybe they'll let us sleep. These exotic new pets, that is, scratching the carpet, howling and chewing while they slobber through the house. We could lock them in back, but they're the passport that got us here. Besides, the old tenants did that and stunk up the room so bad we had to tear up the floor, build it over. They keep getting bigger. It's a good thing he's coming tomorrow.

THE NEW RECRUITS

The officials promised they'd come, that things would get better, and we've seen it: the quiet patches into the morning, a few messages filtering through with the rations. When we're lucky there's something to eat and time to choke it down. Of course the nights are still tricky: explosions rattle the windows, and that scream—are you all right? Am I? Then I woke and my body clenched, gave up the week. Those boys just laid there and laughed. Just doing their job, everyone says. When we finally hobble back home, there isn't any. The foreign music takes over the night as the party empties the neighborhood. Practice that smile and don't let it slip; maybe they won't notice you weren't invited. The new recruits learn fast. They know who's winning.

BUBBLE-BOGGLED

ANTLER

Under the Locust Street Bridge at midnight
      in the middle of the frozen Milwaukee River
         alone with a bottle of wine,
    the starry nightsky twinkling on either side,
Getting on my knees, kneeling on the snow,
      looking where the wind blew the snow away
         exposing the ice like a window,
    a window I can see through,
A black window I can look through
      putting my face to its surface
         to ogle and be boggled
    by bubbles frozen
         at different levels
    in different shapes and sizes,
         white in color,
    suspended, motionless,
And thinking the moment these bubbles froze
      wondering if anyone ever saw
         the moment a bubble froze,
    the moment an air globule
         gurgling and burbling
    on its upward rush
         caught solid in icy hold.
What goes on in a frozen bubble?
Does a frozen bubble believe

it will still be a frozen bubble
 after it melts?
Thought of when they melt,
 rising at last, freed. . . .
Thought of people who drowned
 whose last bubble breaths
 froze midway,
 frozen last words waiting for spring
 and those who listen for them. . . .
Thought of bubbles lasting millions of years
 in icecaps. . . .
Thought of bubbles trapped in lava,
 dark airpockets in rock aeons. . . .
Thought of bubbles rising from canoe paddles
 unstuck from swamp muck. . . .
Bubbles in puddles created and destroyed
 by falling rain. . . .
Bubbles with rainbows quivering
 at the base of waterfalls. . . .
Hippopotamus fartbubbles big as hulahoops,
 frogfartbubbles small as a needle's eye. . . .
Thought of underwater spiders who struggle bubbles of air
 to their underwater webs to breathe from. . . .
Thought of bubbles of thought in cartoons. . . .
Thought of bubbles sparkling up bottles
 stared at by drunks for centuries. . . .
Thought of carpenter observing bubble in his level
 as he adjusts the angle of a beam. . . .
Thought of whales in love caressing each other
 with bubbles. . . .
Thought of girls bobbling their baubles
 goggled by bubble-blowing boys. . . .
Thought of babyblubbering hushed by motherbreast,
 bubble of milk on sleeping lips. . . .
Thought of Imagination Bubble-wand dipped in solution
 strewing bubble flotillas on the breeze,
 different sizes and shapes of poems
 at different levels
 rising and frozen as they rise,

        mind-bubbles caught for a moment
            observed suspended in time
       floating, reflecting. . . .
Thought how I'm only a bubble
      rising from birth to death
          changing my shape
     from child to man as I rise. . . .
Thought of the Earth as a bubble,
    the Sun as a bubble,
         the Galaxies bubbles
    sparkling, flowing, bursting
        on the black river of space,
   on the black river of time. . . .
Thought of the sound of a bubble's pop. . . .
Thought how many bubbles there have been. . . .
Everpresent evanescent efferverscence.
Mind-boggled by bubbles
    I gaze with awe
        through black window ice
Realizing bubbles frozen in ice
    as if I never saw them before,
      as if I never knew
  they existed,
Bubbles frozen in ice,
How I bent to look at them,
How I crouched on my hands and knees
    on the snow
And put my face to the ice
    and peered down at them
       motionless, suspended,
   a long time
Milwaukee River New Year's Eve 1984.

THREE POEMS

LÉON-PAUL FARGUE

Translated from the French by Edouard Roditi

SONG

For our use, the makers have displayed
The usual wares, those objects
That we have always preferred. . . .

The sound of crystal aroused
As if from light slumber hasn't
Disturbed, no, never disturbed
People from their prosperity.

They've produced them in great quantities
Without being moved by their beauty
And, to meet the demand of sales,
There's our little sister, the lamp
That watches all our kisses.

Yes, the lamp, our little sister, casts
Her glance all around and sees
Our kisses. She was asleep like the dead,
Silent in a green mound's hollow.

All day long she kept closed on her part
In the cast and concentrated,

Never uttering a sound, like a beehive
That's silent throughout the winter.

But the time has come. A small
Star trembles and falls headlong.
In the windowpane's sad azure
The fly stops its learned chatter

While the lamp spreads its light that's soft
And pale, of the color of beaches
And wheatfields, the color of sands.
Yes, of all the desert's sands.

We know not in what house
The evening creeps upstairs, leaning
On danger's arm, and stops
Before a landing's marked door.

DAWNS

Let the dawn bring the new breeze
And play at puss-in-the-corner
Nostalgically in the cities
At the crossroads adorned with mirrors
That attract from the depths of the solemn
Distances the subtlest old glances.

Let the rats that silently roll
From one tree to the next, past their gates
To the stream growing pale at this hour,
Go beyond your shadow grown taller
While things respond to your glance
As fast as your glance too to them.

And the shops of the butchers where blood
That was sleeping now slowly drips,

Let them open like flowers to this tremulous
Mauve, let the sky with dull sounds
Rise from the nearly soundless river's
Far end, where a tugboat lows and smokes
From its black nostril against the daylight.

Let the baker's apprentice close
The oven where old ashes still sparkle
And let a vigilant woman
With a mother's eyes and a maid's
Beneath a door where the wind rises
Blow out her half-kilned charcoals
As they sing and with slow hands pour black coffee.

Let the dawn tangle the harsh wind
In the tree where the moon combs her hair
And awaken at last the pond
That's coated with a plum's smooth skin
Where the weirdest insects tremble
As sensitive as the scales
On an old and sleepy cloud.

It's enough, for you to sing
A low and lost song that's all
About women and azure returns
To countrysides, promises and poems,
And it's enough for your heart to darken
And weep from shedding ancient tears.

SONG OF THE ONE WHO IS LIGHTER THAN DEATH

At top speed in hot seats
That crystallize in the heights
We cut the fair short! This isn't Montmartre!
And it's not down there
When the canon roars!

Nor is it war
In the lowing parks!
We are the men without battlements!
We rise in a choir in the music!

Each one back to his stand
While the gods parade,
Little gods who recruit
The fire that in space
Confuses all truths!
This way to mysticism
And the true and only
Spiritualist Sanhedrin,
The polypary of schisms
And the schizogenesis
Of the Council of Trent,
The fart of tycoons
And the pace of cannibals,
The massacre of idols
And the blood of Coligny!
This way to the fine arts,
Bach's basalt
And the funeral pyre
Of Wagner, Rembrandt, Michelangelo,
The thunderbolt made flesh!
This way to the thinkers
And the bottles of doctrines,
The pill-boxes of systems
And flasks of hypotheses,
Spirochetes of ideas
All going at top speed
Beneath the burning ice, enough!

NOTES ON CONTRIBUTORS

ANTLER is the author of *Factory* (City Lights) and *Last Words* (Ballantine), and his work has recently appeared in the following anthologies: *Erotic by Nature, Wisconsin Poets Against AIDS, Gay & Lesbian Poetry in Our Time, Nada Poems,* and *Maverick Poets.* Winner of the 1985 Walt Whitman Award and the 1987 Witter Bynner prize, the poet lives in Milwaukee but spends two months each year alone in the wilderness.

Creator of *Streetfare Journal,* a nationwide project of poetry posters for buses and subways, GEORGE EVANS is also the editor of *Charles Olson & Cid Corman Complete Correspondence, 1956–1964* (National Poetry Foundation) and the author of *Nightvision* (poems, from Pig's Press, London).

LÉON-PAUL FARGUE (1876–1947), born in Paris, enjoyed a long and distinguished career as a poet. He traversed various movements, from Symbolism to Cubism, influenced younger writers, and wrote his memoir of life and literary circles in Paris (*Le Piéton de Paris,* 1939). Although EDOUARD RODITI was authorized by Fargue as far back as 1928 to make these translations, it was not until recently that he decided to make a complete volume of them.

JUDY GAHAGAN worked for many years as a psychologist specializing in social psychology and sociolinguistics and has published books on these subjects. More recently, she has worked as an editor for *New Internationalist* magazine and as a freelance journalist. Publishing poetry and fiction in *Critical Quarterly, Ambit,* and *Prospice,* she now spends much of the year in Italy.

Excerpted here are parts of FORREST GANDER's collection of four long poems, *Deeds of Utmost Kindness.* He has just published his first book, *Rush to the Lake* (Alicejamesbooks) and has been the co-editor with C. D. Wright of Lost Roads Publishers for the past nine years.

PAUL HOOVER's novel, *Saigon, Illinois*, was recently published by Vintage. He lives in Chicago and edits *New American Writing*, a biannual literary journal. His books include *Idea* (Figures) and *Nervous Songs* (L'Epervier Press).

Born in Gifu, central Japan, in 1915, the son of a maker of Buddhist altars, NOBUO KOJIMA ranks at the forefront of Japanese novelists writing today. Although not widely translated into English, Kojima is recognized as one of the finest writers of the group labeled by Japanese critics as the *daisan no shinju*—the "third [wave] of newcomers" who began to make literary names for themselves in the early 1950s. A teacher of Japanese at the University of Hawaii at Hilo, LAWRENCE ROGERS' translations have appeared in *ND51, Japan Quarterly, Translation,* and *Monumenta Nipponica*.

LAURA MARELLO's stories have appeared in *Quarterly West, The PEN Short Story Collection, Sonora Review, Mississippi Review,* and *Chicago Review*. "Catch Me Go Looking" won *Quartely West*'s 1987 annual novella contest.

The Possible, FRED MURATORI's chapbook of poems, was recently published by State Street Poems. The four poems appearing here are taken from *Sartre's Gift;* others from that manuscript have been published in *Gargoyle, Poetry Northwest, Bad Henry Review,* and *Creeping Bent*. Muratori is on the staff of the library at Cornell University.

The renowned and prolific author JOYCE CAROL OATES's work has previously appeared in *ND48, ND49,* and *ND50*. Of her more than forty books of fiction, poetry, and nonfiction, the most recent is the novel *American Appetites* (E. P. Dutton).

CRAIG RAINE is the author of four books of poetry. His opera libretto, *The Electrification of the Soviet Union*, was published in 1986 and performed at Glyndenbourne, Berlin, and Wuppertal. He is also the editor of *A Choice of Kipling's Prose* (1987) and serves as poetry editor at Faber & Faber, London.

An American long resident in Paris, EDOUARD RODITI is an internationally known linguist, scholar, and art critic as well as the author and translator of a considerable number of works of fiction, poetry, criticism, and biography. His critical study *Oscar Wilde* (recently revised and enlarged) and his short story collection *The Delights of Turkey* are both available from New Directions.

The six prose-poems by BARRY SILESKY are part of a now completed manuscript, *The Refugees*. His work has appeared in *North American Review*, *Grand Street*, and *Southwest Review*. He has just finished a novel, *Impossible Weather*, and is now at work on a biography of Lawrence Ferlinghetti, forthcoming from Warner Books. He is also the editor of *ACM (Another Chicago Magazine)*.

CHARLIE SMITH's new book of poems, *Indistinguishable from the Darkness*, has just been published by W. W. Norton. His other books include *Red Roads* (National Poetry Series) and the novels *Canaan* (Simon & Schuster) and *Shine Hawk* (British American Publishing). Forthcoming is his collection of novellas, *Indigo Trilogy*. He lives in New York.

The author of six books (most recently, *A Spool of Blue, New and Selected Poems*, from Scarecrow Press), CAROLYN STOLOFF's poems have appeared in *The New Yorker*, *The Nation*, *Partisan Review*, *Poetry Northwest*, and *Antioch Review*.

Please see the "Translator's Note'" preceding JORGE VALLS's "Eight Poems" for information regarding the poet. LOUIS BOURNE lives in Madrid and, over the course of the last seventeen years, has contributed to these pages many translations of contemporary Spanish-language poets.

ROSMARIE WALDROP is a translator and an award-winning poet. With her husband, the poet Keith Waldrop, she runs Burning Deck

NOTES ON CONTRIBUTORS

Press. She is the author of several books of poetry, a novel (*The Hanky of Pippin's Daughter*, Station Hill Press), and a recent collection of prose-poems (*The Reproduction of Profiles*, New Directions). Please see "Notes on the Poets" following the translations, for information about "Six German Poets."

WILLIAM WEBB graduated from Duke University in 1985 with a B.A. in English and earned an M.A. the following year from Johns Hopkins University in their creative writing program. In 1987 he accepted a Hoyns Fellowship from the University of Virginia and for the past two years has lived in Charlottesville, where he is currently at work on a collection of stories.